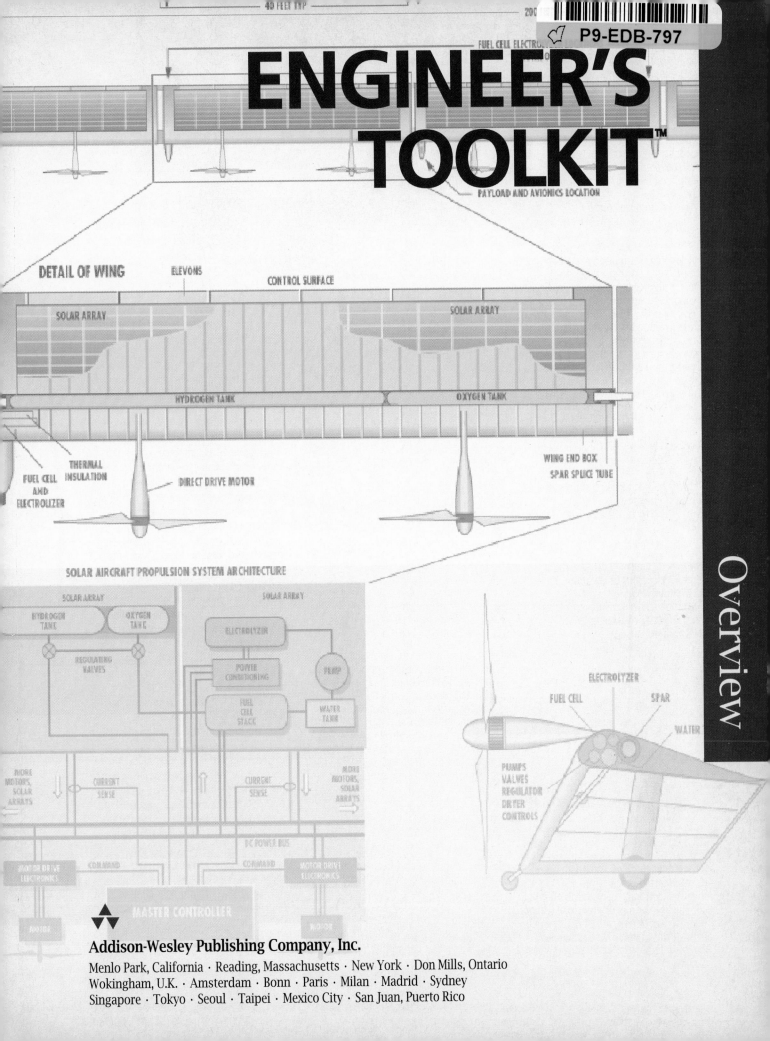

ENGINEER'S TOOLKIT™

FUEL CELL ELECTROL...

PAYLOAD AND AVIONICS LOCATION

DETAIL OF WING

ELEVONS

CONTROL SURFACE

SOLAR ARRAY

SOLAR ARRAY

HYDROGEN TANK

OXYGEN TANK

FUEL CELL AND ELECTROLIZER

THERMAL INSULATION

DIRECT DRIVE MOTOR

WING END BOX

SPAR SPLICE TUBE

SOLAR AIRCRAFT PROPULSION SYSTEM ARCHITECTURE

SOLAR ARRAY

HYDROGEN TANK

OXYGEN TANK

REGULATING VALVES

SOLAR ARRAY

ELECTROLIZER

POWER CONDITIONING

PUMP

FUEL CELL STACK

WATER TANK

MORE MOTORS, SOLAR ARRAYS

CURRENT SENSE

CURRENT SENSE

MORE MOTORS, SOLAR ARRAYS

DC POWER BUS

COMMAND

COMMAND

MOTOR DRIVE ELECTRONICS

MOTOR DRIVE ELECTRONICS

MASTER CONTROLLER

MOTOR

ELECTROLYZER

FUEL CELL

SPAR

WATER

PUMPS
VALVES
REGULATOR
DRYER
CONTROLS

Overview

Addison-Wesley Publishing Company, Inc.

Menlo Park, California · Reading, Massachusetts · New York · Don Mills, Ontario
Wokingham, U.K. · Amsterdam · Bonn · Paris · Milan · Madrid · Sydney
Singapore · Tokyo · Seoul · Taipei · Mexico City · San Juan, Puerto Rico

Executive Editor: Dan Joraanstad
Acquisitions Editor: Denise Penrose
Marketing Manager: Mary Tudor
Developmental Editors: Deborah Craig,
Jeannine Drew, Kate Hoffman,
Shelly Langman
Assistant Editor: Nate McFadden
Senior Production Editor: Teri Holden
Production Editors: Jean Lake,
Gail Carrigan, Catherine Lewis
Supplements Production Editor: Teresa
Thomas
Photo Editor: Lisa Lougee
Copy Editors: Barbara Conway, Robert
Fiske
Proofreader: Holly McLean-Aldis
Marketing Coordinator: Anne Boyd
Cover Design: Yvo Riezebos
Text Design: Side by Side Studios
Technology Support: Craig Johnson
Composition: Side by Side Studios, Fog
Press, London Road Design, Progressive,
The Printed Page
Manufacturing Coordinator: Janet Weaver
Printing and Binding: R. R. Donnelley

Cover and Overview Photo Credits
Cover: Photo ©James Caccavo/Zuma
Images; background illustration
©Ian Worpole
Photo 1: Courtesy of NASA
Photo 2: Courtesy of Jet Propulsion
Laboratory
Photo 3: ©David Parker/SPL/Photo
Researchers, Inc.
Photo 4: ©Brownie Harris/The Stock
Market
Photo 5: Courtesy of M.E. Raichle, Wash.
Univ., St. Louis
Photo 6: ©Chuck O'Rear/Westlight
Photo 7: Courtesy of Lockheed. Photo by
Russ Underwood.
Photo 8: ©Roger Ressmeyer/Starlight
Photo 9: ©George Haling/Photo
Researchers, Inc.
Photo 10: Courtesy of Keith Wood,
Promega, Inc.

The programs, worksheets, and examples
presented in this book have been
included for their instructional value.
They have been tested with care but are
not guaranteed for any particular purpose. The publisher does not offer any
warranties or representations, nor does it
accept any liabilities with respect to the
programs, worksheets, or examples.

ISBN: 0-8053-6335-1

Addison-Wesley Publishing Company, Inc.
2725 Sand Hill Road
Menlo Park, CA 94025

COVER STORY

Pictured on the cover of The Engineer's
Toolkit is the Pathfinder—a "solar-powered
flying wing" designed for low-speed, high-
altitude flight. With a wing span compara-
ble to a Boeing 737, it weighs in at just 400
pounds and has no rudders, no fins, no tail,
and no pilot! The Pathfinder is one of a
series of solar planes developed and built
by Dr. Paul MacCready and his team of
engineers at AeroVironment Inc., in Simi
Valley, California. Engineers at the
Lawrence Livermore National Laboratory in
Livermore, California, designed, engineered, and continue to administer the
Pathfinder solar plane. This laboratory also
is designing the next iteration of solar
planes, the Helios (plans for which appear
behind the photograph of the Pathfinder).
With the Helios, engineers hope to come
even closer to realizing the dream of "eter-
nal flight"; it will include an on-board
energy storage system that can provide the
energy needed during night flight.

As with most contemporary engineering projects, designing solar planes requires
the efforts of engineers from a variety of
disciplines—aeronautical, computer, electrical, environmental, and mechanical, to
name a few. Still other teams of engineers
are needed to design on-board equipment
to support specific missions, such as moni-
toring dangerous weather systems or tracking the release of toxins into the atmosphere.

Contemporary design examples such as
these are presented throughout The Engineer's Toolkit, highlighting the interdisciplinary teamwork that characterizes engineering today.

Tools for a New Curriculum

The Engineer's Toolkit is not a conventional textbook. It was inspired by the needs of instructors like you, who are engaged in developing a new curriculum in introductory engineering courses. They are searching for new ways to prepare, motivate, and engage first-year students. They want to create a link for their students between the prerequisite math and science courses and the wide range of engineering disciplines that build on that knowledge. These instructors also want to ensure that their students master the skills of team-building, communications, and computer use—skills they need to solve problems successfully in subsequent courses and in the real world of work. You and your colleagues are also experimenting with hands-on design projects so students understand that design is a process and that, fundamentally, engineering means solving problems.

Universities and colleges are responding in unique ways to the changing landscape of introductory engineering. This very uniqueness creates a new challenge when you are searching for the right text to support your unique course. *The Engineer's Toolkit* takes on that challenge. You choose from a rich set of course materials that introduce fundamental concepts of engineering and teach essential skills and tools. Each tool is presented as a single module. You determine which modules will best satisfy your course goals, and Addison-Wesley binds those modules into the exact book your students need.

Especially written and designed for *The Engineer's Toolkit,* the modules present a consistent teaching methodology adapted from the work of Delores Etter, author of the spreadsheet and Fortran *Toolkit* modules, as well as *Structured Fortran 77 for Scientists and Engineers.* Each of the *Toolkit* authors has applied

Dr. Etter's five-step problem-solving process to a wide variety of programming languages and application programs. A consistent approach, style, level, and tone means you and your students don't have to switch gears every time you begin to teach or learn a new tool.

Here are the six key pedagogical features you'll find in the *Toolkit* modules that teach programming languages or software tools:

- **The five-step problem-solving process** is explained and illustrated in terms of the particular language or software tool being taught. It is then used throughout the module in applications, numbered examples, and end-of-chapter exercises or problems.
- **Applications** based on the Ten Great Engineering Achievements and representing a wide variety of engineering disciplines demonstrate the five-step problem-solving process.
- **"What If?" problems** immediately follow applications in the software tools modules, asking students to modify assumptions, data, or variables in the application and to solve the new problems that result.
- **Numbered examples** demonstrate key elements of a language or application program by providing fully worked-out solutions.
- **"Try It!" exercises** test students' knowledge of sections within a chapter and frequently require work at the computer.
- **End-of-chapter material** includes summaries of essential points, a key word list, and a set of exercises or problems that gradually increase in complexity.
  These pedagogical features are also described from a student's point of view in the section "The *Toolkit* Methodology."

How To Design Your Custom Textbook

A sample course goal: To introduce engineering, teach a programming language, word processing and CAD techniques.

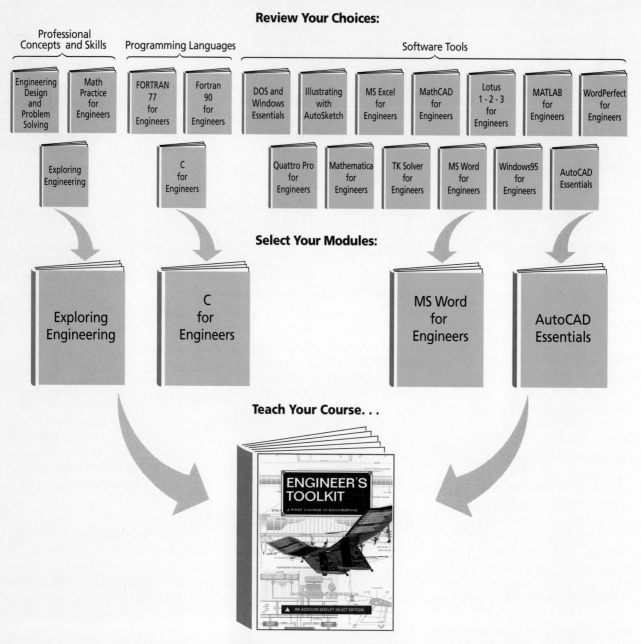

Review Your Choices:

Professional Concepts and Skills | Programming Languages | Software Tools

Engineering Design and Problem Solving | Math Practice for Engineers | FORTRAN 77 for Engineers | Fortran 90 for Engineers | DOS and Windows Essentials | Illustrating with AutoSketch | MS Excel for Engineers | MathCAD for Engineers | Lotus 1-2-3 for Engineers | MATLAB for Engineers | WordPerfect for Engineers

Exploring Engineering | C for Engineers | Quattro Pro for Engineers | Mathematica for Engineers | TK Solver for Engineers | MS Word for Engineers | Windows95 for Engineers | AutoCAD Essentials

Select Your Modules:

Exploring Engineering | C for Engineers | MS Word for Engineers | AutoCAD Essentials

Teach Your Course. . .

ENGINEER'S TOOLKIT
A FIRST COURSE IN ENGINEERING

▲ AN ADDISON-WESLEY SELECT EDITION

With Your Custom Textbook!

DESIGN YOUR OWN TOOLKIT

The *Toolkit* menu consists of three modules that teach programming languages, thirteen modules that teach software tools, and three modules that focus on core engineering concepts and skills—*Engineering Design and Problem Solving, Exploring Engineering,* and *Math Practice for Engineers.* You can mix and match as your course demands.

In a course that combines an overview of engineering disciplines with essential computer applications and the basic design process, you might, for example, combine *Exploring Engineering, Engineering Design and Problem Solving, Quattro Pro for Engineers,* and *Microsoft Word for Engineers* for a book of about 460 pages.

If you plan to teach programming and introduce your students to design and basic CAD techniques, you might combine *Engineering Design and Problem Solving, C for Engineers,* and *AutoCAD Essentials* for a book of about 550 pages.

The *Toolkit* modules are also available separately, and each custom text includes this Overview, which introduces instructors and students to the *Toolkit* methodology.

FUTURE MODULES

Each year Addison-Wesley will release additional modules to keep pace with the development in introductory engineering courses. We welcome your suggestions for future modules in *The Engineer's Toolkit*. Please correspond with us at the Internet address

toolkit@aw.com

or by regular mail at

Toolkit
Addison-Wesley Publishing Company, Inc.
2725 Sand Hill Road
Menlo Park, CA 94025

Toolkit information is also available on the World Wide Web at

http://www.aw.com/cseng/toolkit/

ELECTRONIC SUPPLEMENTS AND SUPPORT MATERIAL

Addison-Wesley offers a range of support materials for *The Engineer's Toolkit*, both printed and electronic. We welcome your comments on the effectiveness of these materials and your suggestions for additional supplements.

Supplements via the Internet A selection of supplementary material for *The Engineer's Toolkit*, including transparency masters, is available on the Internet via anonymous FTP. The URL for accessing this material is ftp://aw.com/cseng/toolkit.

Instructor's Guides Instructor's Guides for the modules you choose will be provided upon adoption of *The Engineer's Toolkit*.

Each Instructor's Guide opens with an overview of the module subject matter, suggested methods of instruction, comments on tests and quizzes, and a general discussion of software version and platform issues (if applicable).

Each Instructor's Guide also describes how to order the module topics to support various syllabi.

A chapter-by-chapter section presents teaching strategies, points to emphasize and special challenges for each chapter, solutions to the end-of-chapter problems, and additional problems and solutions.

Student Data Files Student Data Files, including the data, applications, and program files that support end-of-chapter problems in specific modules, will be available in the following ways:

- ◆ on the disk that includes the Instructor's Guide
- ◆ via our FTP site: ftp://aw.com/cseng/toolkit/igs/
- ◆ via our Toolkit home page on the World Wide Web: http://www.aw.com/cseng/toolkit/

Modules with Data Files include *AutoCAD Essentials, Illustrating with AutoSketch, FORTRAN 77 for Engineers, Fortran 90 for Engineers, C for Engineers, Quattro Pro for Engineers, Lotus 1-2-3 for Engineers, Microsoft Excel for Engineers, TK Solver for Engineers,* and *Mathematica for Engineers.*

Errata for Published Modules: Errata notices for published modules will be available online at: http://www.aw.com/cseng/toolkit/

FIRST ASSIGNMENT

The next section of this Overview is directed to your students. It explains the teaching and learning strategies adopted by our authors throughout the *Toolkit*. The final section introduces students to the Ten Great Engineering Achievements. We invite you to read on and hope you will assign this section to your students early in your course.

THE ENGINEER'S TOOLKIT MODULES		
	Title	**Author**
Professional Concepts and Skills	Engineering Design and Problem Solving	Steve Howell
	Exploring Engineering	Joe King
	Math Practice for Engineers	Joe King
Software Tools	AutoCAD Essentials	Melton Miller
	DOS and Windows Essentials	Gerald Lemay
	Illustrating with AutoSketch	Gordon Snyder
	Lotus 1-2-3 for Engineers	Delores Etter
	MathCAD for Engineers	Joe King
	Mathematica for Engineers	Henry Shapiro
	MATLAB for Engineers	Joe King
	Microsoft Excel for Engineers	Delores Etter
	Microsoft Word for Engineers	Sheryl Sorby
	Quattro Pro for Engineers	Delores Etter
	TK Solver for Engineers	Robert J. Ferguson
	Windows95 for Engineers	Gordon Snyder
	WordPerfect for Engineers	Sheryl Sorby
Programming Languages	C for Engineers	Kenneth Collier
	FORTRAN 77 for Engineers	Delores Etter
	Fortran 90 for Engineers	Delores Etter

The Toolkit Methodology

Welcome to *The Engineer's Toolkit!* This book has been especially created to support your work in what is probably your first course in Engineering. Unlike other textbooks you have studied, *The Engineer's Toolkit* was customized by your instructor to include the exact material you need, and only the material you need. In essence, *The Engineer's Toolkit* is a collection of modules that teach engineering concepts and skills, software tools, and programming languages. Your instructor has selected the appropriate modules for your course, and Addison-Wesley has bound those modules into this custom book.

Introduction. Each application is fully described and explained so that you have sufficient information to complete step 1.

A GENERAL PROCESS FOR SOLVING PROBLEMS

A key feature of *The Engineer's Toolkit* is its emphasis on developing problem-solving skills. Problem solving is one of the foundations of all engineering activity. In *The Engineer's Toolkit* you'll find a five-step method for solving the problems given in each module. Some engineers will tell you they use a nine-step process; others can condense their process down to four. There's nothing magic about the number, but you will find that learning and following a consistent method for solving problems will

make you an efficient student and a promising graduate. Each application program or programming language module builds on this general problem-solving method:

1. Define the problem.
2. Gather information.
3. Generate and evaluate potential solutions.
4. Refine and implement the solution.
5. Verify the solution through testing.

---

66   QUATTRO PRO FOR ENGINEERS

and choosing it. An icon will be on this page for each graph book. Choose the icon of the graph you want to rename with and select Properties|Current Object. The Name dialog box th in which you may type the new name of the graph and then ch

Saving and printing graphs is similar to saving and print sheets. The graph is automatically saved when you save the file. You can print the graph either by itself or with the spre print the graph by itself, select the graph using Graph|Edi choose File|Print. If you have placed the graph as a floating g spreadsheet, it is printed when you print the spreadsheet.

 Try It Change the name of the GRAPH1 graph in the FILTER1 spreads NALIN. Then print the graph.

Application 1   **QUALITY CONTROL**

Manufacturing Engineering
In a manufacturing or assembly plant, quality control receives tion. One of the key responsibilities of a quality control engine lect accurate data on the quality of the product being manufa data can be used to identify the problem areas in the assemb the materials being used in the product.

Circuit Board Defects
In this application, information collected over a one-year peri specify both the type of defects and the number of defects de assembly of printed circuit boards. These defects have been four categories: board errors, chip errors, processing errors, tion errors. Board errors are typically caused by defects in the of the printed circuits. Chip errors are caused by defective in cuit (IC) chips that are added to the board; these IC chips inclu chips, microprocessor chips, and digital filter chips. Processin typically caused by errors in inserting the chips in the board; is often done by manufacturing robots, and the robot progra be incorrect, or the chips being inserted can be packaged i order. Connection errors are solder errors that occur when the through the solder machine; these errors can be caused by board or an incorrect solder temperature.

Spreadsheet for a Quality Analysis Report
You want to develop a spreadsheet that summarizes the qua data that has been collected each month for a year. This data number of defects in each of the four categories of defects dis summary report should compute totals and percentages fo year and defect totals for each quarter. Later in this chapter the data in the spreadsheet to generate pie, bar, and line grap

From Quattro Pro® for Engineers

---

184   C FOR ENGINEERS

```
void init_array(struct sample_type array[MAX_RAINFALLS][MAX_SITES])
{
  int row, col;        /* Loop control variables */

  for (row=0; row < MAX_RAINFALLS; ++row)
    for (col=0; col < MAX_SITES; ++col){
      array[row][col].date.day = 0;
      array[row][col].date.month = 0;
      array[row][col].date.year = 0;
      array[row][col].time.hour = 0;
      array[row][col].time.minutes = 0;
      array[row][col].h_concentration = 0.0;
      array[row][col].ph_level = 0.0;
    }
}
```

Carefully examine the definition of init_array(). Notice that the parameter array is declared as a two-dimensional array of structures. In the body of this function access is made to elements within array using the familiar subscript notation. Once a particular element has been accessed in this manner, the fields of that structure are accessed using the dot operator. Whenever the accessed field is itself a structure, its fields are accessed using a second dot operator, as in the statement

```
array[row][col].date.month = 0;
```

Because the dot is an operator, it can be combined with other operators in this way.

There are many powerful ways to use structures in a program, and this section provides you with only an introduction. In the following application you will learn how structures can be used to extend the mathematical power of C to include complex arithmetic.

Application 1   **FREQUENCY DISTRIBUTION GRAPHING**

Industrial Engineering
Quality-control engineers monitor the quality of an automated production line by tracking the number of defective parts coming off the line within a particular period. If the frequency of defective parts rises dramatically for a given period, the engineer is alerted that a problem exists and can take action to fix the problem. Such frequencies can be depicted using a bar graph such as the one shown in Figure 7-5. In this graph, the horizontal axis represents the number of defects detected and the vertical axis represents data collection periods.

From C for Engineers

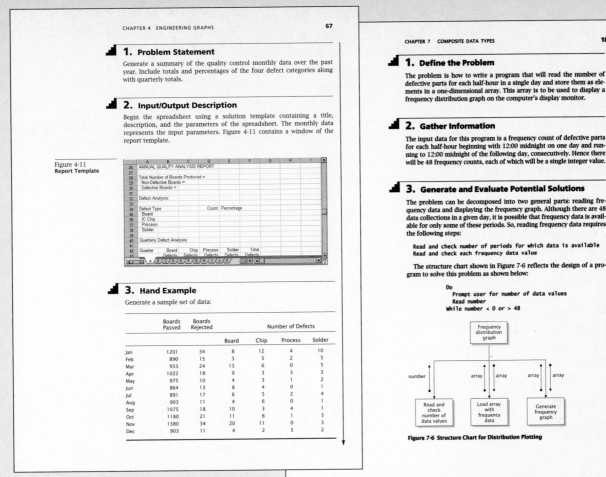

1. Problem Statement

Generate a summary of the quality control monthly data over the past year. Include totals and percentages of the four defect categories along with quarterly totals.

2. Input/Output Description

Begin the spreadsheet using a solution template containing a title, description, and the parameters of the spreadsheet. The monthly data represents the input parameters. Figure 4-11 contains a window of the report template.

Figure 4-11
Report Template

3. Hand Example

Generate a sample set of data:

	Boards Passed	Boards Rejected	Board	Chip	Process	Solder
Jan	1201	34	8	12	4	10
Feb	890	15	3	5	2	5
Mar	933	24	13	6	0	5
Apr	1022	18	9	3	3	3
May	975	10	4	3	1	2
Jun	864	13	8	4	0	1
Jul	891	17	6	5	2	4
Aug	903	11	4	6	0	1
Sep	1075	18	10	3	4	1
Oct	1180	21	11	6	1	3
Nov	1380	34	20	11	0	3
Dec	903	11	4	2	3	2

From *Quattro Pro® for Engineers*

1. Define the Problem

The problem is how to write a program that will read the number of defective parts for each half-hour in a single day and store them as elements in a one-dimensional array. This array is to be used to display a frequency distribution graph on the computer's display monitor.

2. Gather Information

The input data for this program is a frequency count of defective parts for each half-hour beginning with 12:00 midnight on one day and running to 12:00 midnight of the following day, consecutively. Hence there will be 48 frequency counts, each of which will be a single integer value.

3. Generate and Evaluate Potential Solutions

The problem can be decomposed into two general parts: reading frequency data and displaying the frequency graph. Although there are 48 data collections in a given day, it is possible that frequency data is available for only some of these periods. So, reading frequency data requires the following steps:

Read and check number of periods for which data is available
Read and check each frequency data value

The structure chart shown in Figure 7-6 reflects the design of a program to solve this problem as shown below:

Do
    Prompt user for number of data values
    Read number
While number < 0 or > 48

Figure 7-6 Structure Chart for Distribution Plotting

From *C for Engineers*

Step 1. In both applications you are asked to **define the problem.** The introductory description gives lots of clues to help you.

Step 2. This step asks you to **gather information** that you need to propose a solution. In these applications you need to prepare the data that will be used to generate quality analysis reports.

Step 3. You are now ready to **generate and evaluate potential solutions.** In C you create a structure chart to illustrate the design of a program that can solve the problem. In Quattro Pro, the hand example shows how the gathered data will be used to determine the algorithm in step 4.

A SPECIFIC FIVE-STEP PROCESS

Each module adapts this general method and refines it according to the kinds of problems solved by the tool or language being taught. Chapter 1 of each programming language and application program module describes the specific five-step process used in that module.

As you work through *The Engineer's Toolkit*, you'll find that this consistent approach makes it easier to solve new problems. For instance, step 1 of the five-step problem-solving process calls for the same kind of thinking process whether you are using a programming language like Fortran 90, a computer-aided design package like AutoCAD, or an equation solver like MATLAB.

We illustrate the five-step problem-solving process with a pair of applications from *The Engineer's Toolkit*. Both of the examples presented here deal with the collection and tracking of data related to quality

control. Each has been fully worked out using the five-step process. Follow these steps to see how easy it is to learn this problem-solving process.

APPLICATIONS

Step 4. In this step you write a C program based on the structure chart and algorithms developed in the previous steps. You will now **refine and implement the solution.** With Quattro Pro, this step means developing the formulas that will be used to compute the values listed in the summary report.

Step 5. In the final step you **verify the solution through testing.** In C this involves entering a variety of values (value testing) to confirm that the program generates valid output. In Quattro Pro the spreadsheet is tested with several sets of data to verify the accuracy of the computations. Accuracy is confirmed by comparing the spreadsheet calculations to the values determined in the hand example.

These two applications are among hundreds in which *Toolkit* authors demonstrate the five-step problem-solving method. As you gain experience using this method with various software tools and languages, you'll find you can approach new problems with confidence, and you'll begin to identify the appropriate tool or language for the problem at hand. Learning to choose the right tool for a specific engineering problem is an important part of your education.

Many of the applications in *The Engineer's Toolkit* are based on the Ten Great Engineering Achievements chosen by the National Academy of Engineering to celebrate its silver anniversary in 1989. Studying these applications will help you see the kinds of problems faced by engineers from different disciplines and better understand how large problems are broken into smaller solvable problems. This Overview concludes with a description of the Ten Great Engineering Achievements.

---

4. Algorithm Development

The spreadsheet now contains everything except the formulas for computing the summary information for the report. Develop the formulas in the order needed to compute the values. Also, try to minimize the number of computations. For example, since you generate error sums by quarters, add the quarterly sums to get yearly sums instead of adding all the monthly sums to get yearly sums. It would be good practice to verify each of these formulas by referring to Figures 4-10 and 4-11.

D29	@SUM(D11..D22)	Total non-defective boards
D30	@SUM(E11..E22)	Total defective boards
E28	+D29+D30	Total boards
E29	+D29/$E$28	Percent non-defective boards
E30	Copy from E29	Percent defective boards
B44	@SUM(G11..G13)	Quarter I board defects
B45	@SUM(G14..G16)	Quarter II board defects

5. Testing

An important part of developing a spreadsheet is testing it with several sets of data to verify the accuracy of the computations. Using the sample set of data from the hand example, you can easily check the accuracy of the spreadsheet calculations by comparing them to the hand example.

You should make minor changes in this data and check the report to be sure that corresponding changes occur in the report summary. In this report you want to be sure that the report would be generated correctly if there were no errors in one of the categories. Be sure to change the corresponding sums of boards rejected. The corresponding report generated, shown in Figure 4-12, shows that there were no process defects during any of the four quarters.

**Figure 4-12
Report with No
Processing Defects**

ANNUAL QUALITY ANALYSIS REPORT

Total Number of Boards Produced =		12,423
Non-Defective Boards =	12,217	98.34%
Defective Boards =	206	1.66%

Defect Analysis:

Defect Type	Count	Percentage
Board	100	48.54%
IC Chip	66	32.04%
Process	0	0.00%
Solder	40	19.42%

Quarterly Defect Analysis:

Quarter	Board Defects	Chip Defects	Process Defects	Solder Defects	Total Defects
I	24	23	0	20	67
II	21	10	0	6	37
III	20	14	0	6	40
IV	35	19	0	8	62

---

4. Refine and Implement a Solution

The structure chart and algorithms developed in the previous section are implemented as the program in Example 7-10.

5. Verify and Test the Solution

To properly test this program, you should enter a variety of values for first input, including 0, 48, values below 0, values above 48, and valid values. Selecting values along the boundaries is known as boundary value testing. You should do the same for the actual frequency values. Given the input values −3 (error) 55 (error) 10−9 (error) 15 16 13 14 20 9 96 (error) 19 14 19 20, the output of this program is

```
-----------------------------------------------
 1 | ***************
 2 | ****************
 3 | *************
 4 | **************
 5 | ********************
 6 | *********
 7 | ********************
 8 | **************
 9 | *******************
10 | ********************
    --- 5---10---15---20---25---30---35---40---45---50
```

This chapter introduced you to one-, two- and three-dimensional arrays, as well as structures. Arrays and structures offer a cohesive means of storing composite collections of data. Arrays are used for grouping items of the same type and meaning, whereas structures allow you to group related items with different types and meanings. You learned how to declare and initialize arrays and manipulate the elements in an array. You also learned how to define a structure, declare structure variables, and manipulate the fields in a structure. Finally, you learned that these composite data types can be combined to allow you to create arrays of structures, structures containing arrays, and structures containing other structures.

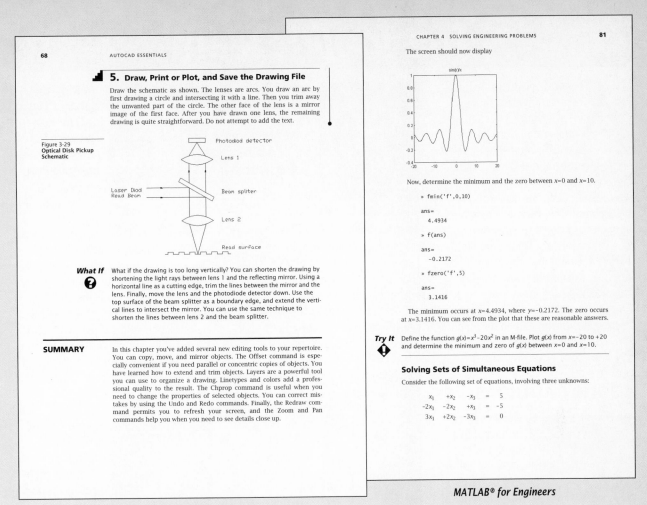

AutoCAD® Essentials

MATLAB® for Engineers

ENGINEERING DISCIPLINES

In *AutoCAD Essentials,* the author presents applications from mechanical and materials engineering, civil and structural engineering, electrical and electronics engineering, and optical engineering. *MATLAB for Engineers* includes applications from electrical and computer engineering. You can find the listing of application problems in Chapter 1 of each software tool or programming language module. The table below lists several applications from the FORTRAN 77 module.

Applications	ACROSS THE DISCIPLINES	
		Chapter
Stride Estimation	Mechanical Engineering	2
Light Pipes	Optical Engineering	3
Sonar Signals	Acoustical Engineering	4
Wind Tunnels	Aerospace Engineering	5
Oil Well Production	Petroleum Engineering	6
Simulation Data	Electrical Engineering	7

❓ "What If" Problems

These problems immediately follow an application. Often you are asked to test the model developed in the application with different data or different assumptions. Working through these problems ensures you fully understand the application.

⬥ "Try It!" Exercises

Each chapter contains several "Try It!" exercises. A "Try It!" is a short set of exercises that tests your understanding of the material. These exercises increase in complexity over the course of each module, and if you try out each one, you'll find you'll master the material faster than you expected.

NUMBERED EXAMPLES

Numbered examples appear in all the programming language and some software tool modules. These examples are designed to illustrate specific features of the language or software. Working through these examples is essential, especially those that offer two solutions and a discussion of the differences between them.

EXERCISES/PROBLEMS

Every module includes end-of-chapter exercises or problems that increase in difficulty and test your knowledge of the chapter material. Make sure you try your hand at the excercises that require you to use the five-step method to find solutions.

The *Toolkit* Team

As you consider the Ten Great Engineering Achievements explored at the end of this Overview, you'll notice that many require contributions from several different engineering disciplines. Today, significant projects can only be accomplished by teams of professionals. And that's true of *the Engineer's Toolkit,* too. The team that has helped to create the *Toolkit* includes not only the authors but also the focus group participants who honed and directed the *Toolkit* concept and the reviewers who helped develop the individual manuscripts.

TOOLKIT AUTHORS

Delores Etter, author of five modules, *FORTRAN 77 for Engineers, Fortran 90 for Engineers, Lotus 1-2-3 for Engineers, QuattroPro for Engineers, and Microsoft Excel for Engineers,* is a professor of electrical and computer engineering at the University of Colorado at Boulder. Dr. Etter has helped shape *The Engineer's Toolkit* from its initial conception, contributing the five-step problem-solving process and key pedagogical features that were successfully tested in her earlier Addison-Wesley texts, such as *Structured FORTRAN 77 for Engineers and Scientists, Lotus 123: A Software Tool for Engineers,* and *Quattro Pro: A Software Tool for Engineers.*

Ken Collier, Assistant Professor of Computer Science in the Department of Computer Science and Engineering at Northern Arizona University in Flagstaff, is the author of *C for Engineers.* Professor Collier teaches courses in C, C++, software engineering, and engineering design. His areas of research include software engineering, software design methodologies, and artificial intelligence.

R. J. Ferguson, the author of *TK Solver for Engineers,* is a professor of mechanical engineering at the Royal Military College of Canada. He teaches courses in stress analysis and computer-aided design. Other publications by Professor Ferguson include texts in the fields of fracture mechanics, noncircular gearing, vehicle transmissions, and engineering education.

Steve Howell, Associate Professor of Engineering in the mechanical engineering department at Northern Arizona University in Flagstaff, Arizona, is the author of *Engineering Design and Problem Solving.* He teaches a course titled Introduction to Engineering Design and Graphics. Professor Howell's areas of research are computer-aided design and manufacturing, thermodynamics, and heat transfer.

Joe King, the author of *Math-CAD for Engineers, MATLAB for Engineers, Exploring Engineering,* and *Math Practice for Engineers,* is an associate professor of electrical engineering at the University of the Pacific in Stockton, California. He teaches courses in electrical engineering, advanced digital design, local area networks, neural networks, machine vision, C++, and microprocessor applications. He conducts research in the areas of neural networks and microprocessor applications.

Gerald Lemay, Professor of Electrical and Computer Engineering at the University of Massachusetts, Dartmouth, is the author of *DOS and Windows Essentials.* He teaches the Science of Engineering for honors students and Computer Tools for Engineers. Professor Lemay does research in renewable energy.

Melton Miller, author of *Auto-CAD Essentials,* is an associate professor of civil engineering and assistant dean of the College of Engineering at the University of Massachusetts, Amherst. He teaches courses in Pascal, Fortran, Lotus, MathCAD, and AutoCAD. He also teaches courses in the the design of reinforced concrete structures.

Henry Shapiro is the author of *Mathematica for Engineers.* He is an associate professor of computer science at the University of New Mexico, where he teaches courses in computer programming and mathematical foundations of computer science. Professor Shapiro conducts research in the area of algorithm design. He is also active in curriculum development and accreditation of computer science programs.

Sheryl Sorby is the author of *WordPerfect for Engineers* and *Microsoft Word for Engineers.* She is an assistant professor of civil and environmental engineering at Michigan Technological University in Houghton, where she teaches courses in freshman engineering and computer skills. Professor Sorby conducts research in structural engineering.

Gordon Snyder, author of *Illustrating with AutoSketch* and *Windows 95 Essentials,* is an associate professor and department co-chair of the departments of electronics systems engineering, computer systems engineering, and laser electro-optics technology at Springfield Technical Community College in Springfield, Massachusetts. He teaches a course titled Introduction to Computer-Aided Engineering Technology.

REVIEWERS AND FOCUS GROUP PARTICIPANTS

Instructors throughout the country attended focus groups to help us identify key trends in engineering education. Over 100 reviewers contributed to the development of manuscripts for *The Engineer's Toolkit.* We gratefully acknowledge all their contributions.

10 Great Engineering Achievements

In celebration of its silver anniversary, the U.S. National Academy of Engineering identified the Ten Great Engineering Achievements accomplished during the organization's first 25 years. Initially selected because they represent major breakthroughs, these achievements have initiated whole new areas of engineering. In the pages that follow, we note the types of design problems faced by interdisciplinary teams of engineers who work in these fields, and we represent contemporary examples in the photographs. Many of the applications you will encounter throughout *The Engineer's Toolkit* are based on these achievements.

1

SPACE TRAVEL

Although 1972 marked the last in a series of moon landings begun in 1969, the Apollo mission laid a foundation for a whole new generation of space shuttle missions that were dedicated to gathering information about the universe and that continue to test the human ability to travel in space.

Design Problems Several key design problems had to be solved to support the Apollo mission to land humans on the moon. The spacecraft required a new inertial navigation system; the lunar lander ascent engine had to work perfectly because there was no backup engine; the spacesuits had to protect the astronauts in a hostile environment and yet be flexible enough to allow movement; and the Saturn V rocket, which powered all the Apollo flights, had to be 15 times more powerful than the biggest rockets available in the early 1960s.

Application Areas Today, rockets launch deep space probes such as the Voyager, which continues to send back images from as far away as Venus and beyond, while space shuttles launch information-gathering satellites and scientific equipment. The Hubble Space Telescope (HST) was launched from the

The space shuttle Endeavour floats above the earth at an altitude of 381 miles, with the west coast of Australia forming the backdrop for the 35mm frame. While perched on top of a foot restraint on the Endeavour's Remote Manipulator System arm, astronauts F. Story Musgrove (top) and Jeffrey A Hoffman wrap up the last of five space walks. They have succeeded in their mission to repair the Hubble Space Telescope.

space shuttle Discovery on April 25, 1990. NASA had designed HST to allow scientists to view the universe up to 10 times more sharply than they could with earthbound telescopes. Unfortunately, scientists soon discovered that the primary mirror in HST was flawed and could not focus properly. It was another space shuttle, the Endeavour, that operated as a kind of Hubble repair station.

Engineering Disciplines Although the fully operational Hubble Space Telescope now stands as an impressive achievement for NASA, the repair job itself was perhaps an equally important achievement. Materials engineers helped develop metals, plastics, and other materials that could withstand the rigors of the launch and the environment in space. Mechanical engineers helped develop the mechanical structures that position HSTs mirrors and other moving apparatus and electrical engineers helped develop the complex computer, communication, and power systems.

2

APPLICATION SATELLITES

Application satellites and other spacecraft orbit the earth to capture, relay, and transmit specific types of information, or to perform manufacturing processes that rely on special properties of the extraterrestrial environment, such as zero gravity.

Application Areas Satellite systems provide information on weather systems, relay communication signals around the globe, survey the earth and outer space to map uncharted terrain and provide navigational information for vehicles on land, in the oceans, and in the air. The Endeavour has been involved in NASA's Mission to Planet Earth, which is designed to help the international scientific community better understand which environmental changes are caused by nature and which are induced by human activity. Throughout 1994 the shuttle orbited the earth with the Spaceborne Imaging Radar-C and X-Band Synthetic Aperture Radar system, which illuminates the earth with microwaves, allowing detailed observations at any time, regardless of weather or sunlight conditions. (See photo below.)

Design Problems A satellite or spacecraft launching system must be designed to generate enough thrust to escape the earth's atmosphere. Once free, it needs to maintain a stable orbit around the earth. In addition, the hull needs to be light and yet strong enough to withstand the stress of the liftoff.

Engineering Disciplines Space-based inventions, such as the Spaceborne Imaging Radar-C (SIR), and satellites in general are a result of the cooperative efforts of aerospace engineers who help develop the systems that put satellites in space, and of chemical, mechanical, and electrical engineers who assist in the development of the radar and imaging systems for applications such as SIR.

3

MICROPROCESSORS

A microprocessor is a tiny computer, smaller than your fingernail, that combines the control, arithmetic, and logic functions of large digital computers.

Application Areas With its small size and powerful capabilities, the

A technician working on the Bit Serial Optical Computer (BSOC), an optical computer that stores and manipulates data and instructions as pulses of light. To enable this, the designers (Harry Jordan and Vincent Heuring at the University of Colorado) developed bit-serial architecture. Each binary digit is represented by a pulse of infrared laser light 4 meters long. The pulses circulate sequentially through a tightly wound 4-kilometer loop of optical fiber some 50,000 times per second. Other laser beams operate lithium niobate optical switches which perform data processing.

applications of microprocessors range from operating remote television controllers or VCR recorders to providing the computational power in hand-held calculators or personal computers. Microprocessors can also be found in communication devices, such as networks that connect computers around the globe, and in automobiles, ships, and airplanes.

Design Problems Key design problems involved in creating microprocessors include miniaturization, increasing speed while controlling the heat produced, and searching for materials stable and reliable enough to store, process, and transmit data at high speeds. For decades engineers have improved the performance of computers by increasing the number of functions contained on the CPU chip. Ultimately this approach created a bottleneck in switching between these functions. In the early 1980s, chip designers addressed this issue and developed a concept of Reduced Instruction Set Computing (RISC) which improves efficiency through the high-

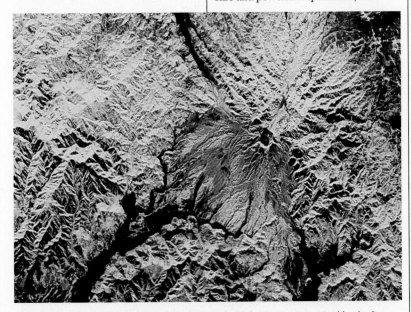

This image of the area around Mount Pinatubo in the Philippines was acquired by the Spaceborne Imaging Radar-C and X-Band Synthetic Aperture Radar system aboard the space shuttle Endeavour in April 1994. This false color image shows the main volcanic crater on Mount Pinatubo produced by the June 1991 eruptions and the steep slopes on the upper flanks of the volcano. The red color shows the rougher ash deposited during the eruption. The dark drainages are the smooth mudflows that continue to flood the river valleys after heavy rains. This radar image helps identify the areas flooded by mudflows, which are difficult to distinguish visually, and assess the rate at which the erosion and deposition continue.

speed composition of an optimized minimal set of instructions. Research and development is on-going in another area of microprocessor design as well: optical computing. A dream since the 1940s, optical computing represents a fundamental change in how switching occurs—through optical signals rather than electronic signals. Since a computer is nothing more than a complex system of on/off switches, the speed at which the switches can turn off and on is the single most critical factor in determining the computer's performance. Engineers can design optical switches that operate well into giga-hertz range, while electronic switches are currently restricted to about 100 megahertz.

Engineering Disciplines An application such as optical computing is a result of many years of collaboration between electrical engineers who

Inspection of the largest, most powerful energy-efficient jet engine, which was designed for the Boeing 777 jets.

design the complex laser systems, computer engineers who work on the computational structures, and chemical engineers who develop the actual lasers.

4

JUMBO JET

Much of the success of jumbo jets (747, DC-10, L-1011) can be attributed to high-bypass fanjet engines, which allow the planes to fly farther using less fuel and with less noise than previous jet engines. Jumbo jets also have an increased emphasis on safety. For example, a 747 has four main landing-gear legs instead of two; a middle spar was added to the wings in the event one is damaged; and redundant hydraulic systems operate the critical system of elevators, stabilizers, and flaps that control the motion of the plane.

The newest jumbo jet is Boeing's 777, which was entirely developed using computer-aided design systems.

Application Areas The technological advances of superior fuels, engines, and materials achieved during the creation of the jumbo jet benefited the space program and the military.

Design Problems A major design problem faced by creators of the jumbo jet was creating an engine and fuel supply that could generate sufficient horsepower and thrust needed to lift the huge aircraft off the ground. The hull had to be strong enough to withstand the stress of the flight, without being so heavy that fuel efficiency was lost. The internal environment had to be comfortable for the passengers, especially during liftoff and landing.

Engineering Disciplines Engineers have been developing increasingly powerful jet engines ever since Englishman Sir Frank Whittle developed the first jet engine prototypes in 1937. The evolution of Whittle's engine into the Boeing 777 engine was largely due to the efforts of mechanical engineers who specialized in dynamics, thermodynamics, combustion systems, and materials.

5

MEDICAL IMAGING

A CAT (computerized axial tomography) scanner is a machine that generates three-dimensional images or slices of an object using x-rays. A series of x-rays is generated from many angles, encircling the object or patient. Each x-ray measures a density at its angle, and by combining these density measurements using sophisticated computer algorithms, an image can be recon-

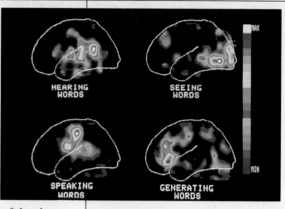

The PET scans shown here reveal localized brain activity under four different conditions, all related to language. Physicians use the PET scanner to diagnose brain and heart disorders and certain types of cancer.

structed that gives clear, detailed pictures of the inside of the object. A PET (positron-emission tomography) scanner reveals locations of intense chemical activity within the body. Sugar labeled with radioactive isotopes that emit particles called positrons is first injected into a person's bloodstream. These positrons collide with electrons made available by chemical reactions in the body. The scanner detects the energy released by these collisions and maps metabolic "hot spots"—regions of an organ that are most chemically active at the time.

Application Areas Doctors use diagnostic radiology to both detect and diagnose diseases, and in combination with other medical techniques, to treat diseases. These devices can also be used to explore the structure of both organic and inorganic materials.

Design Problems The key design problem lies in getting a clear picture without harming the patient or technician. In addition, the machine needs to withstand the various fields it produces.

Engineering Disciplines The PET and similar medical scanning systems are a product of biomedical engineering, a very specialized field of engineering that combines electrical engineering with medicine. Also involved are materials engineers—mechanical engineers who develop the special structures that contain, direct, and withstand the sometimes hazardous electromagnetic and radioactive emissions often used in such systems.

6

ADVANCED COMPOSITE MATERIALS

A composite consists of a matrix of one material that has been reinforced by the fibers or particles of another material. The choice of the composite is determined by the need of the application. For example, does the application require a material that is strong, flexible, stiff, lightweight, heavy, heat-tolerant, porous, dense, or wear-resistant?

Application Areas Advanced composites can be found in most products, including automobile components, communication systems, building materials, artificial joints and organs, and machine parts.

Design Problems Composite material designers must determine how the various materials will interact with each other and how the composites will act over time given the often high-stress situations in which they are used.

Engineering Disciplines Mechanical engineers are most frequently involved in the development of composite materials. Working with chemists and sometimes chemical engineers, mechanical engineers design applications, such as artificial limbs, for particular composite materials.

7

CAD/CAM/CAE

CAD (computer-aided design) refers to computer systems used by engineers to model their designs. These systems may include plotters, computer graphics displays, and 2-D and 3-D modelers. CAM (computer-aided manufacturing) systems are used to control the machinery or industrial robots used in manufacturing the parts, assembling components, and moving them to the desired locations. CAE (computer-aided engineering) systems support conceptual design by synthesizing alternative prototypes using rule-based systems; they also support design verification through rule checking of CAD models. CAD/CAM/CAE systems are intended to increase productivity by optimizing design and production steps, and by increasing flexibility and efficiency.

Application Areas CAD/CAM/CAE systems are used in all engineering design disciplines to support product design, test, and production.

Design Problems Engineers must correctly convert real-world specifications to valid computer models, design appropriate computer tests for the model, and design computer systems to accurately and economically manufacture the final product.

Engineering Disciplines CAM has had a major effect on the work of many industrial engineers. Switching over from labor-intensive manufacturing to computer-aided manufacturing has changed and will continue to change the national and international industrial environ-

Zina Bethune, ballet teacher, has special long artificial hip implants coated with cobalt and chrome. She needed the implants because she suffers from degenerative arthritis.

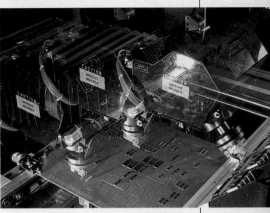

This automated integrated circuit insertion device, designed by Lockheed, is installing an integrated circuit into a circuit board for a satellite program.

ment. Because they use CAM systems themselves, industrial engineers are heavily involved in their development. Computer and electrical engineers design the computer and control systems and write the software that drive CAM systems.

8

LASERS

Light waves from a laser (that is, light amplification of electromagnetic waves by stimulated emission of radiation) have the same frequency and thus create a beam with one characteristic color. More importantly, the light waves travel in phase, forming a narrow beam that can easily be directed and focused.

Application Areas CO_2 lasers can be used to drill holes in materials such as ceramics, composite materials, and rubber. Medical uses of lasers include repairing detached retinas, sealing leaky blood vessels, vaporizing brain tumors, removing warts and cysts, and performing delicate inner-ear surgery. Lasers are used in scanning devices to scan Universal Product Codes. High-power lasers also are used in weapons and to create 3-D holograms.

Design Problems A key concern is controlling the heat and power generated by lasers so the laser does not harm the patient or the product.

Engineering Disciplines Although they were not heavily involved in the original design and development of lasers in the 1950s, engineers have developed applications for them in

Automatic analysis of DNA is performed with laser beams by Leroy Hood and Jane Sanders, biologists at California Institute of Technology. This DNA sequenator is also called the "Gene Machine."

many areas: Electrical engineers use them in optical fiber communications, civil engineers use them to perform accurate surveying, and mechanical engineers use them for precise cutting of metal parts.

9

FIBER OPTIC COMMUNICATIONS

An optical fiber is a transparent thread of glass or other optically transparent material. It can carry more information than either radio waves or electrical waves in copper telephone wires. In addition, fiber optic communication signals do not produce electromagnetic waves that cause cross-talk noise on communication lines. The first transoceanic fiber optic cable was laid in 1988 across the Atlantic. It contains four fibers that, together, can handle up to 40,000 calls at one time.

Application Areas An increasing number of communications and computer systems are converting to fiber optics due to its enormous information capacity, small size, light weight, and freedom from interference.

Design Problems A key design problem is to create cables that can withstand the stress of being buried under ground or under the ocean. Another design problem is the need to

keep the various signals independent and free from outside interference.

Engineering Disciplines Mechanical engineers design the manufacturing systems that produce glass and plastic optical fibers. Electrical engineers design the transmitters, amplifiers, and receivers that, along with the fibers, carry the optical signals from source to destination.

10

GENETIC ENGINEERING

A genetically engineered product is created by splicing a gene that produces a valuable substance from one organism and placing it into another organism that will multiply itself and the foreign gene along with it. Genes are artificially recombined in a test tube, inserted into a virus or bacteria, and then inserted into a host organism in which they can multiply. Once the new organism has been created, a system has to be designed to produce and process the product in large quantities at a reasonable cost.

The first commercial product of genetic engineering was human insulin, which appeared commercially under the trade name Humulin. The molecules are produced by the genetically engineered bacteria and are then crystallized into human insulin.

Application Areas In addition to creating new drugs and vaccines, genetic engineering has been used to create bacteria that can clean up oil

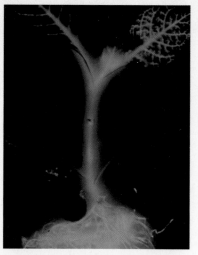

Researchers incorporated firefly gene codes for the enzyme that catalyzes the chemical reaction to release energy in the form of light, into the DNA of a tobacco plant.

spills and detoxify wastes. The process is also used to create genetically altered plants that are pest and disease resistant or have certain desirable characteristics such as improved taste, shipping hardiness, or longer shelf life.

Design Problems Engineers must translate the laboratory work of biologists into the large-scale commercial manufacturing systems that are robust, safe, and cost effective.

Engineering Disciplines Mechanical engineers design equipment for growing large quantities of genetically engineered organisms, chemical engineers design processes for separating out the desired end product, and environmental engineers evaluate the potential impact on the environment.

Thinner than a human hair, an optical fiber can carry more information than conventional radio waves or electrical waves.

Microsoft Excel for Engineers

Delores M. Etter
Department of Electrical and Computer Engineering
University of Colorado, Boulder

Addison-Wesley Publishing Company

Menlo Park, California · Reading, Massachusetts · New York
Don Mills, Ontario · Harlow, UK · Amsterdam · Bonn
Singapore · Paris · Tokyo · Madrid · San Juan, Puerto Rico
Milan · Seoul · Taipei · Mexico City

The worksheets presented in this book have been
included for their instructional value. They have been
tested with care but are not guaranteed for any particular purpose. The publisher does not offer warranties or representations, nor does it accept any
liabilities with respect to the examples.

Microsoft is a registered trademark of Microsoft Corporation. Some of the product names used herein
have been used for identification purposes only and
may be trademarks of their respective companies.

This is a module in The Engineer's Toolkit™, an
Addison-Wesley SELECT™ edition. The Engineer's
Toolkit and SELECT are trademarks of the Addison-
Wesley Publishing Company. Contact your sales rep-
resentative for more information.

Photo Credits:
Chapter 1: ©John Madere/The Stock Market
Chapter 2: ©Courtesy of NASA Wallops Flight Facility
Chapter 3: ©Todd Powell/Profiles West
Chapter 4: Courtesy of Intel™
Chapter 5: ©Carl Purscell/Photo Researchers, Inc.

ISBN 0-8053-6535-4

Addison-Wesley Publishing Company
2725 Sand Hill Road
Menlo Park, CA 94025

Contents

Chapter 1: **Problem-Solving with Microsoft Excel   1**

Introduction   2
1-1   **Solving Problems with Microsoft Excel   2**
1-2   **Using a Five-Step Problem-Solving Process   3**
Summary   6
References   6

Chapter 2: **An Introduction to Microsoft Excel   9**

Introduction   10
2-1   **Using Worksheets   10**
Exploring the Excel Window   11
Moving Within the Worksheet   13
Entering Values in the Worksheet   14
Entering Text in the Worksheet   15
Editing Data   15
2-2   **Working with Commands   16**
The Menu System   16
Dialog Boxes   17
Toolbar Buttons   17
2-3   **Performing Common Worksheet Operations   19**
Selecting a Range of Cells   19
Formatting the Worksheet   20
Manipulating Columns and Rows   21
Copying and Moving Information   22
Formatting with Toolbar Buttons   22
Printing a Worksheet   22
Saving and Opening a Worksheet   23
2-4   **Accessing Online Help   24**
Application 1   **Sounding Rocket Trajectory—Aerospace Engineering   25**
Rocket Stages   25
Worksheet Containing Altitude Data   25
Summary   26
Exercises   27

Chapter 3: Engineering and Scientific Computations 29

Introduction 30

3-1 Creating Mathematical Formulas 30

Cell References 30

The Formula Bar 31

Formula Operations 31

Copying Formulas 33

Recalculation and Iteration 33

Circular References 34

3-2 Using Mathematical Functions 35

Common Functions 35

Trigonometric Functions 36

Exponential and Logarithmic Functions 37

Rounding and Truncating Functions 38

3-3 Including Statistical Computations 38

Statistical Functions 39

Range Names 40

Application 1 **Random Number Generation—Electrical/Computer Engineering 41**

Bounds for Random Numbers 41

Worksheet for Generating Uniform Random Numbers 41

3-4 Using Special Functions 43

IF Function 43

LOOKUP Functions 44

DATE and TIME Functions 46

Application 2 **Cable Car Velocity—Mechanical Engineering 47**

Cable Car Velocity Equations 47

Worksheet for Generating Cable Car Velocity Values 47

Application 3 **Temperature Distribution--Mechanical Engineering 50**

Thermal Equilibrium 50

Worksheet for Computing Temperature Distribution 51

3-5 Creating Macro Commands 54

Summary 55

Exercises 56

Chapter 4: Engineering Graphs 59

Introduction 60

4-1 Creating *XY* Charts 60

Creating a Simple Chart 61

Changing the Range 64

Enhancing a Chart 64

Naming, Saving, and Printing a Chart 66

Application 1 **Quality Control—Manufacturing Engineering 66**

Circuit Board Defects 66

Worksheet for a Quality Analysis Report 67

4-2 Creating Other Types of Charts 71

Pie Charts 71

Column Charts 73

Line Charts 75

4-3 Importing Data Files 77

Importing Files 78

Parsing Imported Data   80
Summary   80
Exercises   80

Chapter 5:   **Scientific Databases**   **83**

Introduction   84

5-1   **Creating a Database Using an Excel List**   **84**
Specifying Fields and Records   84
Viewing Titles   86
Displaying Record Numbers   86

Application 1   **Climatology Data—Environmental Engineering**   **87**
Creating a Worksheet for a Climatology Database   88
Descriptive Information   88
Number of Fields   89
Types of Data   89
Field Names   91

5-2   **Sorting a List**   **92**
Sort Keys   92
Sorting Orders   93

5-3   **Updating a List**   **95**
Inserting New Fields and New Records   96
Filtering Lists   96
Modifying List Records   102

5-4   **Performing Statistical Computations**   **103**
Database Functions   104
The Histogram Tool   106

Summary   108
Exercises   108

Index   **111**

1 Problem-Solving with Microsoft Excel

Computer-Aided Engineering

Engineers use computers to support the design, testing, and manufacturing processes. The graphics software in a computer-aided design (CAD) system helps engineers create computer models or software prototypes based on initial design specifications. A computer-aided engineering (CAE) system helps refine the model through testing, and a computer-aided manufacturing (CAM) system directs machines, such as robotic arms, to assemble the final product. Although computers streamline engineering design and analysis, using them effectively requires problem-solving skills at every step of the process.

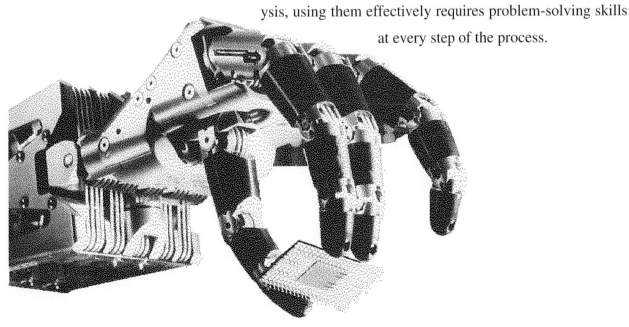

INTRODUCTION

In this chapter you will develop problem-solving skills by using a five-step problem-solving process. This chapter introduces you to the spreadsheet program Microsoft Excel, examines the types of problems a spreadsheet program is designed to solve, and then presents a five-step process for solving a wide variety of problems. Finally, a simple example that computes the average of a set of laboratory measurements demonstrates this five-step process with Microsoft Excel.

1-1 SOLVING PROBLEMS WITH MICROSOFT EXCEL

A *computer* is a machine designed to perform operations that are specified with a set of instructions called a *program*. Computer *hardware* refers to the computer equipment, and computer *software* refers to the programs that describe the steps you want the computer to perform. One way to solve a problem with the computer is to write a program in a *programming language*, which provides a set of instructions that can be used to tell a computer what to do. For example, Fortran and C are two programming languages used by engineers and scientists. Today, however, for some classes of problems, you can save time by using powerful *application programs*—programs designed to perform specific tasks and functions. Spreadsheet programs are a type of application program that use an electronic worksheet divided into rows and columns; you can then use these programs to analyze, calculate, and present data from the *spreadsheet* or *worksheet*.

The first spreadsheet program, VisiCalc, was developed in 1978 by Dan Bricklin, a student at the Harvard Business school, and his friend Bob Frankston, a student at MIT. The idea for VisiCalc came to Bricklin as he was analyzing financial-planning problems that required numerous recalculations whenever a financial assumption was changed. Bricklin felt the repetitive recalculations could be performed more easily by a computer, and Frankston turned Bricklin's idea into reality. The program Frankston wrote was for the Apple II computer, and VisiCalc's instant popularity added to Apple's success.

When the IBM PC was introduced in 1981, an IBM version of VisiCalc quickly followed, as did the development of several competing spreadsheets. In 1981 Mitch Kapor developed the Lotus Development Corporation. In 1983 Kapor's company introduced Lotus 1-2-3, which was to become one of the most successful application programs ever. This book focuses on Microsoft Excel, which is considered to be an *integrated package*—a package that incorporates several functions. You can perform calculations on your data using mathematics and statistics functions, display your data as a graph using graphics functions, quickly locate data using database functions, and document your spreadsheet using word processing functions. Microsoft Excel is available in both DOS and Windows versions. The version this module focuses on is Microsoft Excel Release 5.

Microsoft Excel provides over 200 engineering, math, statistics, and logic worksheet functions you can use to perform calculations needed to solve most engineering and science problems. For example, you can use the predefined functions to calculate trigonometric and logarithmic values, analyze a linear regression, or determine if a condition is true or false.

If the function and calculation capabilities provided by Excel are not suited to solving a specific problem, you might want to consider using a specialized mathematical package, such as MathCAD or MATLAB. Writing a program in a programming language such as Fortran or C is another option for solving the problem. However, with each new release, the power and features of Excel are enhanced, and the types of problems you can solve using the software continues to grow.

1-2 USING A FIVE-STEP PROBLEM-SOLVING PROCESS

Problem solving is an everyday activity. Problems range from analyzing lab data to finding the quickest route to work. While computers can solve many problems, a computer tool or a computer language is most effective when a problem is broken into steps that the computer can perform. The problem-solving process introduced in this text is based on a general five-step problem-solving process. In the following table this general procedure is shown to the left, and the procedure you will use in this module is shown to the right.

A General Problem-Solving Model	Excel Problem-Solving Model
1. Define the problem.	Problem statement.
2. Gather information.	Input/output description.
3. Generate and evaluate potential solutions.	Hand example.
4. Refine and implement a solution.	Algorithm development.
5. Verify and test the solution.	Testing.

Understanding the five steps is easier when you apply them to a familiar problem, such as computing an average. This problem arises, for example, when you compute homework averages or summarize data from a lab experiment.

Step 1 of the process is to define the problem. It is important to give a clear, concise statement of the problem to avoid any misunderstandings. For this example the problem statement is

> Compute the average of a set of experimental data values.

Step 2 consists of gathering information. You need to describe carefully any information or data needed to solve the problem and then describe how to present the final answer. These two types of information are called input and output, respectively. Collectively they are called *input/output (I/O)*. In this example the input information is the list of experimental data values. The output is the average of these data values.

Step 3 is to generate and evaluate potential solutions. In this step you work the problem by hand or with a calculator, using a simple set of data. An example of a set of data values and the computed average is

Lab Measurements—1/4/95

Number	Value
1	23.43
2	37.43
3	34.91
4	28.37
5	30.62
	154.76

Average = Sum/5 = 30.95

A simple problem such as this may have only one possible solution, and identifying the solution may take you only a few minutes. More complex problems may have several solutions, and generating these solutions and determining which is most suitable can take much longer.

Step 4 is to refine and implement a solution. In this step you describe, in general terms, the operations you performed by hand. The sequence of operations that solves the problem is called an *algorithm*. The procedure that we use in algorithm development in this module is called *top-down design*. Top-down design consists of two techniques: *decomposition* and *stepwise refinement*. The problem is broken into a series of smaller problems (decomposition), and each smaller problem is addressed separately (stepwise refinement). Decomposition is a form of "divide and conquer," in which each part of the overall problem is described in general terms. Stepwise refinement begins with this general description and successively refines and describes each step in greater detail. The refining continues until the solution is specific enough to convert into the language of the computer tool or into a programming language.

An advantage of decomposition is that it allows you to think of the overall steps required to solve the problem without getting lost in the details. Details are introduced only as the refinement of the algorithm begins. Programming languages like Fortran and C use block diagrams and pseudocode or flowcharts to present the algorithm steps in a form that can be translated into the language. A software tool like Excel uses a series of computations in the worksheet to solve the problem. It is important that the algorithm computations be decomposed into steps that relate to the capabilities of the tool or language that is going to be used to solve the problem. For example, when computing an average with Fortran or C, the program should add the values and divide by the number of values. Microsoft Excel computes the average with a single function reference; therefore, once you enter the data in the worksheet, you just enter a formula using the AVERAGE function to compute this average.

As you develop the algorithm, it is important to use a consistent method of documentation. A *solution template* provides an outline for all worksheet solutions presented in this module. This template includes four components: a title for the worksheet, a description of the problem being solved, a list of the input (or variable) parameters for the solution, and the computations and analyses performed by the worksheet. The worksheet in Figure 1-1 contains a window of the worksheet that computes the average for the hand example and contains all four elements of the solution template.

Figure 1-1
**Worksheet for
Computing an Average**

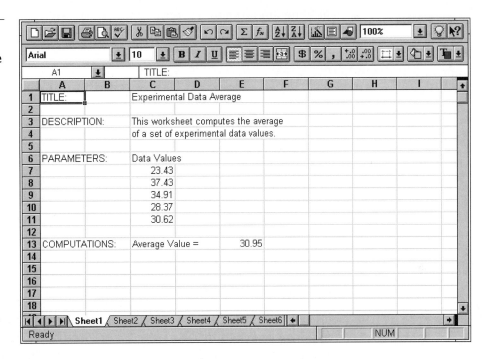

Step 5 in the problem-solving process is to verify and test the solution. Testing an algorithm is not easy. While it is usually possible to find data for which the algorithm works correctly, it is just as important to determine if any sequences of data cause the algorithm to fail. A correct algorithm for averaging data should work properly for any set of data.

The five-step problem-solving process is demonstrated throughout this module in applications. The disciplines of the applications are as follows:

Applications	Across the Disciplines	
Applications	**Discipline**	**Chapter**
Sounding Rocket Trajectory	Aerospace Engineering	2
Random Number Generation	Electrical/Computer Engineering	3
Cable Car Velocity	Mechanical Engineering	3
Temperature Distribution	Mechanical Engineering	3
Quality Control	Manufacturing Engineering	4
Climatology Data	Environmental Engineering	5

SUMMARY

It is important to begin your investigation of engineering and science with a solid methodology for solving problems and then to learn how to use the computer to help solve the types of problems you will encounter. This chapter described Microsoft Excel as an applications program that easily and efficiently performs calculations commonly used for solving engineering and scientific problems. A five-step procedure for developing problem solutions (algorithms) was presented. The five steps are the following:

1. Define the problem clearly.
2. Describe the input and output.
3. Work a simple example by hand.
4. Develop an algorithm in worksheet format.
5. Test the worksheet with a variety of data.

You develop the solution using top-down design. Decomposition assists in describing the general steps that have to be performed to solve the problem. Stepwise refinement guides you in refining the steps and adding necessary details. The chapters that follow show how to use Microsoft Excel to solve a range of interesting problems.

Key Words

algorithm	programming language
application program	software
computer	solution template
decomposition	spreadsheet
hardware	stepwise refinement
input/output (I/O)	top-down design
integrated package	worksheet
program	

References

Many of the outstanding engineering and scientific achievements of the last 25 years have used computer-aided engineering. For more information, see the following references:

MOON LANDING

Hallion, Richard P., and Tom D. Crouch, eds. "Apollo: Ten Years Since Tranquility Base." Washington D.C.: Smithsonian Institution Press, 1979.
Stix, Gary, ed. "Moon Lander." *Spectrum*, Vol. 25, No. 11, 1988, pp. 76-82.

APPLICATION SATELLITES

Canby, Thomas Y. "Satellites That Serve Us." *National Geographic*, September 1983, pp. 281-334.
Heckman, Joanne. "Ready, Set, GOES: Weather Eyes for the 21st Century." *Space World*, July 1987, pp. 23-26.

MICROPROCESSORS

Garetz, Mark, "Evolution of the Microprocessor: An Informal History." *BYTE*, September 1985, pp. 209–215.

Greenberg, Donald P. "Computers and Architecture." *Scientific American*, February 1991, pp. 104–109.

Toong, Hou-Min D. "Microprocessors." *Scientific American*, Vol. 237, No. 3, 1977, pp. 146–151.

COMPUTER-AIDED DESIGN AND MANUFACTURING

Loeffelholz, Suzanne. "CAD/CAM Comes of Age." *Financial World*, October 18, 1988, pp. 38–40.

Mitchell, Larry D. "Computer-Aided Design and Manufacturing." *McGraw-Hill Encyclopedia of Science & Technology*. New York: McGraw-Hill, 1987.

CAT SCAN

Andreasen, Nancy C. "Brain Imaging: Applications in Psychiatry." *Science*, March 18, 1988, pp. 1381–1388.

Sochurek, Howard. "Medicine's New Vision." *National Geographic*, January 1987, pp. 2–40.

ADVANCED COMPOSITE MATERIALS

Chou, Tsu-Wei, Roy L. McCollough, and R. Byron Pipes, "Composites." *Scientific American*, October 1986, pp. 192–203.

Steinberg, Morris A. "Materials for Aerospace." *Scientific American*, October 1986, pp. 67–72.

JUMBO JETS

Ingells, Douglas J. *747: Story of the Boeing Super Jet.* Fallbrook, Calif: Aero Publishers, 1970.

Stewart, Stanley. *Flying the Big Jets.* New York: Arco Publishing, 1985.

LASERS

Berns, Michael W. "Laser Surgery." *Scientific American*, June 1991, pp. 84–90.

Jewell, Jack L., James P. Harbison, and Axel Scherer. "Microlasers." *Scientific American*, November 1991, pp. 86–94.

"Lasers Then and Now." *Physics Today*, October 1988 (Special Issue).

Townes, Charles H. "Harnessing Light." *Science 84*, November 1984, pp. 153–155.

FIBER-OPTIC COMMUNICATIONS

Bell, Trudy, ed., "Fiber Optics." *Spectrum*, Vol. 25, No. 11, 1988, pp. 97–102.

Desurvire, Emmanuelf. "Lightwave Communications: The Fifth Generation." *Scientific American*, January 1992, pp. 114–121.

Drexhage, Martin G. and Cornelius T. Monihan. "Infrared Optical Fibers." *Scientific American*, November 1988, pp. 110–116.

Lucky, Robert W. "Message by Light Wave." *Science 85*, November 1985, pp. 112–113.

GENETICALLY ENGINEERED PRODUCTS

Barton, John H. "Patenting Life." *Scientific American*, March 1991, pp. 40–46.

Bugg, Charles E., William M. Carson, and John A. Montgomery. "Drugs by Design." *Scientific American*, December 1993, pp. 92–101.

Eskow, Dennis, "Here Come the Assembly-Line Genes." *Popular Mechanics*, March 1983, pp. 92–96.

Weaver, Robert F. "Beyond Supermouse: Changing Life's Genetic Blueprint." *National Geographic*, December 1984, pp. 818–847.

2 An Introduction to Microsoft Excel

Sounding Rocket Trajectory Engineers and scientists often use sounding rockets to study the upper regions of the earth's atmosphere. A telemetry system mounted on the nose of the rocket transmits scientific data to a receiver at the launch site as the rocket passes through the different levels of the atmosphere. In addition to the scientific data, performance measurements on the rocket itself are transmitted and monitored by test-range safety personnel and then analyzed later by engineers. One of those measurements is the rocket's altitude from the time of launch until the test is completed. In this chapter you examine an Excel worksheet that contains the rocket's altitude data.

INTRODUCTION

In this chapter you explore the Microsoft Excel window, learning how to use Excel commands and toolbar buttons to initiate operations commonly used with worksheets. These operations include enhancing the appearance of data, manipulating columns and rows, opening and saving files, and printing worksheets. You also discover how to access online Help to find out more about Excel as you use it.

Since learning to use Excel is very much a hands-on experience, whenever possible you should read this module while sitting in front of a computer so you can follow the examples yourself. Short exercises appear frequently in the module, allowing you to see quickly if you have understood the important material just presented; they are called "Try It!" exercises. The more of these exercises that you attempt on the computer, the quicker you will learn Excel. It is not assumed that you have used Excel before, but it is assumed that you are familiar with using a microcomputer. If you need to install Excel on your computer, refer to the *Microsoft Excel User's Guide* for initialization and installation information.

2-1 USING WORKSHEETS

Microsoft Excel stores data and performs computations within a worksheet composed of a rectangular grid of rows and columns. (This worksheet is often called a spreadsheet when used for business accounting and record keeping. Retaining the term worksheet emphasizes the fact that the worksheet can be used in a very wide range of applications.) In this section you discover how to move within the worksheet and how to enter and edit worksheet data. Figure 2-1 shows a blank worksheet.

Figure 2-1
Blank Worksheet

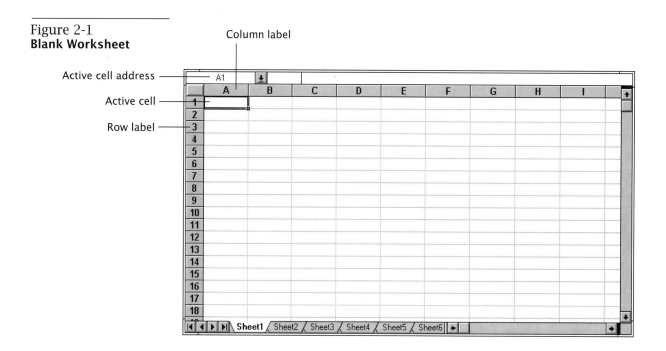

Each column of the worksheet is referenced by a letter (or letters) that range from A through IV; that is, the first group of columns are referenced by A through Z, the next group of columns are referenced by AA through AZ, the next group of columns are referenced by BA through BZ, and so on, with the last group of columns referenced by IA through IV. The worksheet has a total of 256 columns.

Each row is referenced by a number that can range from 1 to 16384. Notice the letters on the columns and the numbers on the rows in Figure 2-1.

Each location, or cell, within the worksheet can be referenced with a *cell address* composed of a column letter and a row number. In Figure 2-1 the cell address of the upper-left cell is A1; the cell address of the upper-right cell is I1; the cell address of the lower-right cell is I18; and the cell address of the lower-left cell is A18. The four corner cell addresses for the largest possible Excel worksheet are A1, IV1, IV16384, and A16384.

Over three million cells are potentially available in a worksheet, although you typically use only a small part of this total. The maximum-size worksheet that you can use is also limited by the amount of memory in your computer system.

Exploring the Excel Window

When you start Excel you initially see the *Excel window*, which contains a *workbook* named Book1 with a blank worksheet identified with a tab Sheet1, as shown in Figure 2-2.

Figure 2-2
Excel Window and Worksheet Window

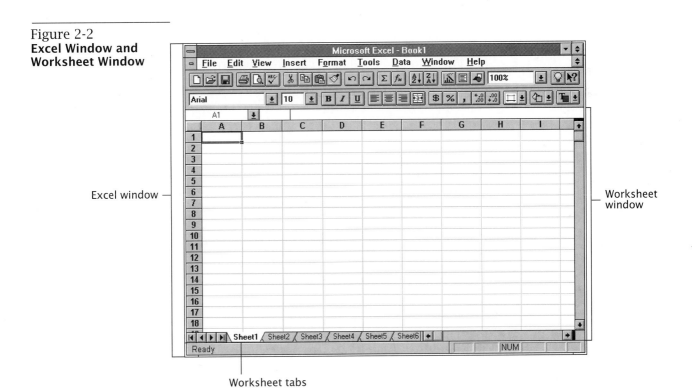

Excel window

Worksheet window

Worksheet tabs

Since only a portion of a large worksheet can be displayed in the window at one time, the window typically displays only 18 rows. The number of columns displayed varies, depending on the width of the columns. Later in this chapter you will learn how to display different portions of large worksheets in the window.

Although Excel allows you to work with more than one worksheet in a workbook, you will only be working with workbooks containing a single worksheet in this module.

Figure 2-3
Control Panel

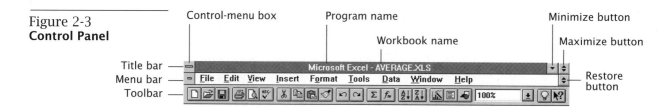

The Control panel, located at the top of the Excel window, displays information about Excel and about the workbook and contains the *title bar*, the *menu bar*, and *toolbars*, as shown in Figure 2-3. You use the Control-menu box to close the Excel window, and use the Minimize, Maximize, and Restore buttons to change the size of the Excel window.

The Title bar displays the program and workbook names.

The Menu bar displays a series of command categories. For example, File includes a set of commands used to manipulate files. Some of the displayed categories change depending on what you are doing.

Toolbars contain buttons that provide shortcuts for many Excel tasks, such as saving or printing your worksheet or undoing a previous operation. The standard toolbar is shown in Figure 2-3; the formatting toolbar is shown in Figure 2-4.

Figure 2-4
Edit Line

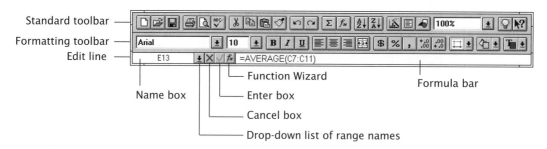

The Edit line, shown in Figure 2-4, includes the following:

The Name box displays the cell address of the data.

The arrow to the right of the Name box can be used to display a list of range names as discussed in Chapter 3.

The Cancel box is used to cancel the current operation.

The Enter box is used to enter the data.

The Function Wizard button is used to access functions to use formulas, as discussed in Chapter 3.

The Formula bar displays cell data and provides you with information, such as numbers, text, functions, and formulas.

Figure 2-5
Worksheet Window

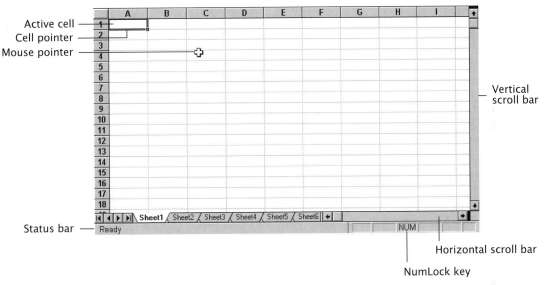

The Status bar, shown in Figure 2-5, is located at the bottom of the Excel window and provides information about the current selection, the current Excel operation, and toggle keys, such as NumLock. You can also use online Help (discussed later in this chapter) to find out more about the status bar.

Moving Within the Worksheet

Now that you are familiar with the various parts of the Excel window, you can begin to move around within the worksheet. When you begin Excel you are located at cell A1. Since the *cell pointer* highlights the *active cell*, cell A1 should be highlighted when you begin working with a worksheet. You can move the cell pointer with the mouse or the arrow keys. The cell pointer, active cell, and mouse pointer are shown in Figure 2-5. To move the cell pointer rapidly to see data in cells not currently in view, either use the mouse to move the scroll bars or hold down one of the arrow keys on the keyboard. Whichever method you choose, the active cell is always highlighted by the cell pointer, and its address is displayed in the left corner of the edit line.

Table 2-1   Moving the Selector

Desired Move	Keystroke
Left or right one column	← or →
Up or down one row	↑ or ↓
Up or down one full screen	PGUP or PGDN
To beginning of column	CTRL+↑
To cell A1	CTRL HOME
To beginning of row	HOME
To cell E2	F5 E2 ENTER
To the lower-right corner of the active area	END HOME

Table 2-1 lists several other ways to move from one cell to another. For instance, pressing PGUP or PGDN moves the cell pointer up or down a full screen at a time. To move to a specific cell, you can press the F5 (GOTO) function key and then enter the cell address. Finally, you can always return to cell A1 by pressing CTRL HOME. (Press both keys at the same time.) You can practice using these options in the following exercises.

Try It

1. Use the mouse or arrow keys to move to the following cells: D36, BC18, E3. Observe the changes in the edit line as you move from cell to cell.

2. Now, using the mouse or arrow keys, scroll to the following cells: GA500, F27, Z176, E2.

3. Use the F5 function key to move to the following cells: GA500, F27, Z176, E2.

Entering Values in the Worksheet

Now that you can move around easily within the worksheet, you are ready to enter data into some of the cells. A cell can contain either *values* (numbers or formulas) or text entries. You need to move the cell pointer to the desired cell before you can enter data in it. To enter a value, type the value and press ENTER to store the value in the cell.

To demonstrate entering a value, move the cell pointer to A1, and then type 2.5 and press ENTER. You should see the value displayed in the cell address and in the *formula bar* (which is also called the *cell contents* box) on the edit line as you type 2.5, as shown in Figure 2-6.

Figure 2-6
Entering Data in the Worksheet

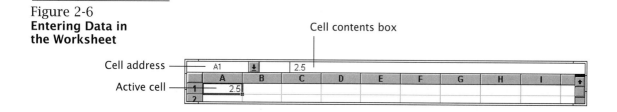

Cell contents box

Cell address

Active cell

If you make an error in typing the value, and if you have not already pressed (ENTER), you can press (BKSP) to erase the error. If you have already pressed (ENTER), you can return to the cell and type the value again.

You must follow a number of rules when you enter numeric values:

- Values must begin with a number, decimal point, plus sign, minus sign, dollar sign, or left parenthesis.
- Values must not include spaces.
- Values may end with a percent sign.
- Values may be entered in scientific format, in which the number is expressed as a number times a power of 10 (for example, the scientific format for 0.000003 is 3.0×10^{-6}, which is written as 3.0E–6).

Numeric values can range between 10^{-4931} and 10^{4932}, but Excel can only display numbers between 10^{-307} and 10^{99}. Thus Excel can handle the range of values for most engineering and scientific applications. Later in the chapter you learn various ways of displaying this information in the worksheet.

You can enter data by columns, with the second value entered below the first value, and so on. The way you can do this is to position the cell pointer at the top of the column, enter a value, and press ⊕. The new active cell is automatically the one in the next position in the column.

Try It

Enter the following values in the indicated cells:

- 0.425 in cell A6
- $5.25 in cell G25
- –2 in cell B289
- 6.2% in cell B299 (note the change in decimal position)
- 5000000 in cell A15
- Enter the numbers 1 through 10 in cells D1 through D10.

Entering Text in the Worksheet

Text refers to information that is not used in numeric operations. Mathematical operations cannot be performed with text, even if the text contains numbers. The initial alignment is determined by a default setting that specifies that the text is left-aligned (or left-justified).

Editing Data

Once you have entered data in a worksheet, you may want to change an entry in a cell. You can completely replace the contents of the cell using the same procedure that you used initially to put data in the cell—simply type the new data and press (ENTER). To edit the entry, point to the cell that contains the entry that you want to edit and double-click the left mouse button, or press the (F2) (EDIT) function key to go into Edit mode. Notice that the contents of the active cell are displayed in the formula bar. You can then press ⓐ and ⓑ to move the insertion point (blinking position) within the active cell. Press (BKSP) to erase characters to the left of the insertion point, and press (DEL) to erase the character at the insertion point. You can also insert text at the insertion point. You may also use the (HOME) and (END) keys to go to the beginning and end of the contents of a cell. When you have edited the entry as desired, press (ENTER) or click the Enter box to store the new value in the cell. Note that the *mode indicator* changes from Ready to Edit and then back to Ready as you update the contents of the cell.

Try It

Enter the text 1/15/95 in cell C5. Then edit it so that it contains

- ◆ 01/15/95
- ◆ 01-15-95
- ◆ Jan 15, 1995
- ◆ JAN. 15, 1995

2-2 WORKING WITH COMMANDS

Microsoft Excel contains over 500 commands that can be used with worksheets. While this large number of commands may seem overwhelming, remember that most people use only a small subset of the commands. At the same time it is this wealth of commands that makes Excel so powerful; there are commands for almost any operation that you want to perform. This module will not cover all the commands. Instead, it presents the general command structure so that you understand the overall categories of commands and how to access them. The next section presents the most commonly used worksheet commands. The later chapters introduce new commands as needed to solve engineering and scientific problems. Refer to the *Microsoft Excel User's Guide* and online Help (discussed later in this chapter) for information on commands not covered in this module.

The Menu System

Microsoft Excel commands are divided into the following categories: File, Edit, View, Insert, Format, Tools, Data, Window, and Help. These menu bar commands are located at the top level of a deeply nested command structure. As the entry point to the command structure, each menu bar command provides you with various options from which you can select a different operation. The options for each menu bar command are organized into a series of pull-down and cascade menus and *dialog boxes*. Figure 2-7 illustrates this menu system.

Menu bar commands are easy to access. To choose a command from the main menu, first point to the command and click the left mouse button. When you choose a command, the status bar displays a description of the command.

Figure 2-7
**Pull-down and
Cascade Menus**

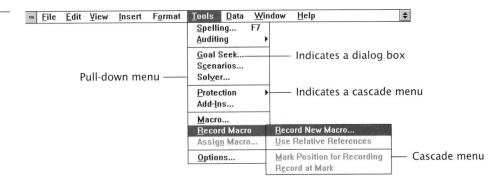

A **pull-down menu** then appears, listing additional commands you can choose. To choose one of the commands from a pull-down menu, point to the command and click the left mouse button. Commands not followed by an arrowhead or an ellipsis (...) are performed immediately. If you choose a command that is followed by an arrowhead, then a cascade menu appears.

The cascade menu provides you with additional command choices. If you choose a command that is followed by an ellipsis, then a dialog box appears.

A **dialog box** allows you to select options and specify data needed for the command. To return to the worksheet without performing a command, click the Cancel button in the dialog box.

Shortcut menus are used by clicking the right mouse button. To use the shortcut menu on worksheet data, first select the data and then click the right mouse button to see the shortcut menu. Shortcut menus contain only the most frequently used Excel commands.

Dialog Boxes

You use dialog boxes to complete Excel tasks. Dialog boxes involve one or more of the following operations: entering text in a text box, selecting a check box, clicking buttons, and selecting an item from a list or drop-down box. When you click OK the operation is performed. You can cancel the command by clicking the Cancel button. Figures 2-8a and 2-8b show typical dialog boxes that include many of these options. You will learn how to use dialog boxes once you begin entering data into worksheets.

Toolbar Buttons

Toolbar buttons provide shortcuts you can use to perform common operations. There are 13 toolbars, each with a set of buttons. Some buttons appear on multiple toolbars. You select a toolbar button by pointing to the symbol (or icon) and clicking the left mouse button. Figure 2-9 shows the standard toolbar. It contains toolbar buttons for frequently used File, Edit, Help, and Insert menu bar commands, along with a few other often-used commands for summing, sorting, and creating charts. Many of these buttons and commands are discussed in later sections.

Figure 2-8a
Dialog Box Options

Tabs

List box

Command buttons

Text box

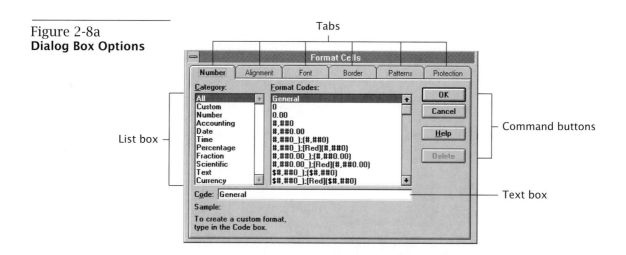

Figure 2-8b
**More Dialog Box
Options**

Option buttons

Check box

Drop-down list

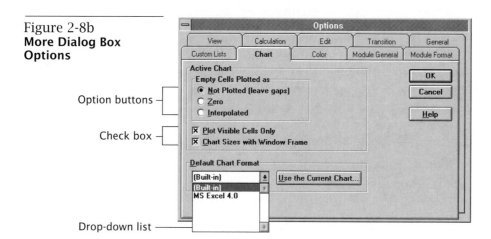

Figure 2-9
Standard ToolBar

Spelling
Print preview
Cut
Copy
Paste
Format Painter
Sort ascending
Sort descending
Zoom Control
Function
Autosum
Drawing
Help
Print
Save workbook
Open workbook
New workbook
Repeat
button
Undo
button
Text box
Chart Wizard
Tip Wizard

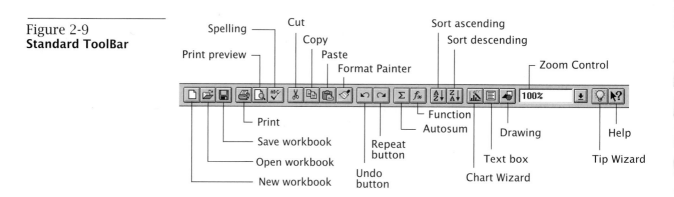

Try It Display each of the following:

♦ A cascade menu by choosing Tools Record Macro. Then cancel the command.

♦ A dialog box by choosing Tools Spelling. Then close the dialog box and pull-down menu.

♦ A description of a toolbar button by pointing to the toolbar button.

2-3 PERFORMING COMMON WORKSHEET OPERATIONS

The most common worksheet operations involve formatting the work-sheet, copying and moving information, printing a worksheet, and saving and retrieving a worksheet. You will use these operations to examine a worksheet that contains altitude data from a sounding rocket trajectory.

Selecting a Range of Cells

Many of the commands covered in this section require that you select the cells you want to use with the command. Excel refers to the selected cells as a range.

A range is a rectangular block of cells, as shown in Figure 2-10, that can be as small as a single cell or as large as the entire worksheet.

A range address is composed of the addresses of the first and last cells in the range separated by a colon (for example, D7:D15 or B2:C15). You can select multiple ranges that may or may not be adjacent or may even over-lap.

To select a single cell, just click the cell. To select a range of cells, first move the cell pointer to the cell in one corner of the range. Then press and hold the left mouse button. While continuing to hold down the mouse button, drag across the worksheet until the range you want is highlighted. If you make a mistake, just click anywhere in your worksheet and start over.

Figure 2-10
Range Address B2:C15

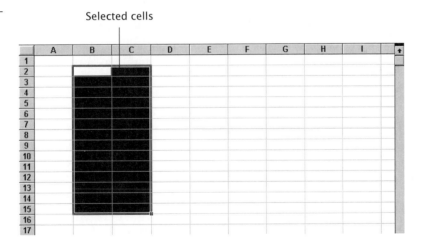

To select multiple ranges at one time, select the first range, and then hold down (CTRL) while you select the next range. Repeat the process until all of the ranges in the collection are selected.

To select an entire column or row, click the column letter or row num-ber. To select an entire worksheet, click the blank button just above the column of row numbers.

Formatting the Worksheet

Excel's commands provide several methods you can use to improve the appearance of your worksheet. For example, you can style numbers as currency or percentages, underline or center text, change the width of columns, or add or delete rows and columns.

Formatting Numbers   You use the Format command to change the form in which numbers, dates, times, and text are displayed in the worksheet; the internal values that are stored in the cells are not affected. The Format Style command works on a single cell, a range of cells, multiple ranges, or the entire worksheet. The following are some of the most commonly used number formats.

Format	Display	Example
Number	Specifies a fixed number of decimal places in the value, with or without a comma	3,002.56
Scientific	Specifies a scientific or exponential format for the value	0.52E+03
Currency	Specifies various currency styles using the dollar sign, commas, and two decimal places for the value or rounding to the next dollar value.	$762.50
Percentage	Specifies a percent format	5.25%
Date	Specifies various styles for displaying a number as a date	12-Dec-95
Time	Specifies various styles for displaying a number as a time	11:59 AM
Fraction	Specifies the value as a mixed fraction	2/3

All numbers with decimals are displayed with up to nine decimal places, unless you specify a smaller amount. If a fixed number is too wide to fit in a cell, it is displayed in scientific format. If a number has too many digits or characters (such as plus, minus, dollar sign, comma, period, and percent) to fit in the width of the cell, Excel replaces the value with pound signs (#). The actual content of the cell is not lost; it just cannot be displayed with the specified format. Later in this section you will learn how to make the cell wide enough to display the number.

To format a range, you first select it. You then choose the Format Cells command. The Format Cells dialog box shown in Figure 2-11 appears. Next select a category from the Category list box and a format code from the Format Codes list box (0 indicates a position that will contain a digit and # indicates a position that will contain a digit or a space).

Aligning Data   By choosing the Alignment tab from the Format Cells dialog box, you can change the way information is positioned in a cell. You can horizontally position information on the left, center, or right of the cell. You also can vertically position information on the top, center, or

Figure 2-11
Format Cells Dialog Box

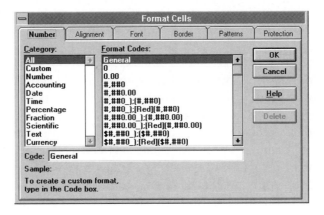

bottom of the cell. You can fill the cell with a repeating pattern. Remember to select the range before choosing the command.

Try It

Enter the numbers 10000, 1000, 100, 10, 1, 0.1, 0.01, 0.001, and 0.0001 in cells D1 through D9. Change the styles to the following specifications:

◆ Fixed with four decimal places (use the Custom format option)

◆ Scientific with one decimal place

◆ Currency

◆ Comma with one decimal place

◆ Percent with one decimal place

◆ Centered

Now enter text data in cells E1 through E9. Then

◆ Center it

◆ Right align it

◆ Fill the contents of E9 with asterisks (*)

Manipulating Columns and Rows

The default column-width setting is eight positions, which may not always be appropriate for the information in your worksheet. You can change the column width for selected cells by choosing Format Column Width.

As you develop your worksheet, you may decide to insert or delete rows and columns. For example, you may want to insert a column of data between two columns that already contain data, or you may want to enter blank rows or columns to separate information and improve the appearance of the worksheet. To delete rows and/or columns, first select the range you want to delete. Remember, you select an entire row or column by clicking the row number or column letter. Adjacent cells shift up and to the left after you select Edit Delete. Rows are inserted above the selected range, and columns are inserted to the left of the selected range when you select Edit/Insert. Adjacent cells shift down or to the right.

Copying and Moving Information

You use the Edit command to copy or move information from one range to another. Use the command Edit Copy to automatically copy the contents of a specified range to the *clipboard*, a temporary storage area. Next select the destination range where you want the information copied; you only need to select the upper-left cell of the destination where you want the information to be copied. Then choose Edit Paste to copy the contents of the clipboard to the destination range.

Moving information is very similar to copying information, except that the information is actually moved instead of copied. Therefore, the data is deleted from the original cells during a move. The only difference between the steps for moving information and copying information is that you choose Edit Cut for a move instead of Edit Copy. You can delete a range of information by using Edit Cut without pasting the contents of the clipboard back into your worksheet.

Formatting with Toolbar Buttons

You also can use toolbar buttons to change styles, alignment, and order. After selecting the range, point to the appropriate toolbar button, and click the left mouse button. The toolbar buttons used to emphasize data by displaying it as **bold**, *italic*, or underlined, or aligned to the left, center, or right are shown in Figure 2-12. As worksheets are developed in later chapters, formatting will be used to improve the readability of the information they contain, particularly numeric information. You can use the Format Painter button on the standard toolbar (see Figure 2-9) to copy the format of one range to another range. First select a cell or range with the formats you want to copy, and then select the Format Painter and drag across the new range.

Figure 2-12
Formatting Toolbar

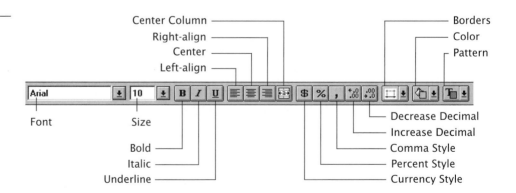

Printing a Worksheet

Excel provides numerous features you can use to print a range of cells or an entire worksheet. When you choose File Print, the Print dialog box shown in Figure 2-13 appears. You may choose to print a selection (if you already selected a range of cells), selected sheets from the workbook, or the entire workbook by clicking the appropriate option button. To preview what the printed pages will look like, select the Print Preview command button. The selection, worksheet, or workbook will then be displayed in the Print Preview window. You close the window by clicking the Close button. To print, click the OK button. You also use the Print button and the Print Preview button on the standard toolbar.

Figure 2-13
Print Dialog Box

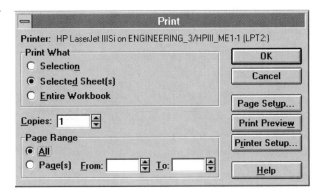

Saving and Opening a Worksheet

After creating a workbook, you should save it for later use. The first time you save a workbook in a file, use the File Save command. The Save As dialog box appears. The default directory is displayed under Directories; BOOK1, the default file name, is displayed with the extension .XLS. To replace the default file name, select it and type a new file name; the workbook will be saved with this file reference and extension. To change the directory where your file will be saved, click the desired directory icon. To change the drive, select the desired drive in the Drives menu. Excel will automatically add the extension. The next time you save the file, the Save As dialog box will not appear. You can use File Save As to save a copy of the file under a different name.

To open a new file, use the File New command. Excel opens a new file using a default file name. To retrieve an existing file, use the File Open command. You specify the file you want to open by typing its name under File Name in the File Open dialog box. Or you can use the File Name and Directories list boxes to select the file you want. You can also open and save files with standard toolbar buttons.

Try It

Enter the numbers 1 through 10 in cells D1 through D10. Save the worksheet using the name WORK1. Then replace the numbers in cells D1 through D10 with zeros. Now retrieve the workbook and observe that the original values are back in cells D1 through D10.

2-4 ACCESSING ONLINE HELP

Online Help provides information about Excel features, such as menu commands, toolbar buttons, and dialog boxes. You can either choose the Help command or press (F1) to access online Help. To find out about a specific Excel feature, choose Search on the pull-down menu. You then use the Search dialog box shown in Figure 2-14 to locate the desired information.

Figure 2-14
Search Dialog Box

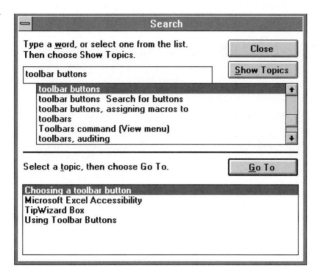

When you first open the Search dialog box, the insertion point is in the text box in the upper portion of the dialog box. Begin typing a word or phrase you want to learn more about. As you type, a list of related Help key words is displayed. Select the appropriate key word by pointing to the word and choosing Show Topics. A list of Help topics then appears. If you point to the appropriate Help topic and choose Go To, Excel displays information about the topic you selected. You can obtain additional information about any green or underlined text by pointing to the word and clicking the left mouse button.

Online Help also provides information about working with the various Excel dialog boxes. To find out how to use a dialog box that is displayed on-screen, just click the Help command button in the dialog box. You close Help by choosing the File Exit menu.

Try It

◆ Use online Help to find out how to use the Open File dialog box.

◆ Use online Help to identify which toolbar buttons display in various formats. (You will need to select the Toolbar Buttons topic.)

◆ Use online Help to identify which toolbar buttons insert and delete rows and columns.

◆ Use online Help to find out about the Undo command and the Repeat toolbar button. These features are very useful because they allow you to undo and repeat most actions. Return to the worksheet and store 99 in cells B3 and C3. Then store 3.0 in cell B3. Assume that storing 3.0 in cell B3 is a mistake, undo the operation, and restore the value of 99 in cell B3. Change the format of cell B3 to currency. Now repeat this command on cell C3.

The next section presents a worksheet that contains altitude information from a sounding rocket trajectory. This worksheet enables you to experiment with the worksheet commands discussed in this chapter.

Application 1	SOUNDING ROCKET TRAJECTORY

Aerospace Engineering

Sounding rockets are used in aerospace engineering to probe different levels of the atmosphere. The altitude data presented in the worksheet in this section is from a two-stage rocket that was launched to perform high-altitude atmospheric research on the ionosphere.

Rocket Stages

The first stage of the rocket burns for approximately 35 seconds and accelerates the rocket to a velocity of 1250 meters per second. It then coasts for almost 2 minutes before reaching the lower region of the ionosphere at about 100 kilometers. By then gravity has slowed the rocket's ascent to about 100 meters per second. The second stage ignites and accelerates the rocket through the ionosphere and into space.

Worksheet Containing Altitude Data

Choose the File Open command to retrieve the TRAJECTORY workbook. (This workbook and all other workbooks from the "Application" sections in this module are available on a diskette that is available to your instructor. Check with your instructor to see if this information has been stored on your computer system so that you can access these workbooks.) Once you have successfully retrieved the file, the window shown in Figure 2-15 appears on your screen. The worksheet shows corresponding time and altitude measurements for a sounding rocket trajectory.

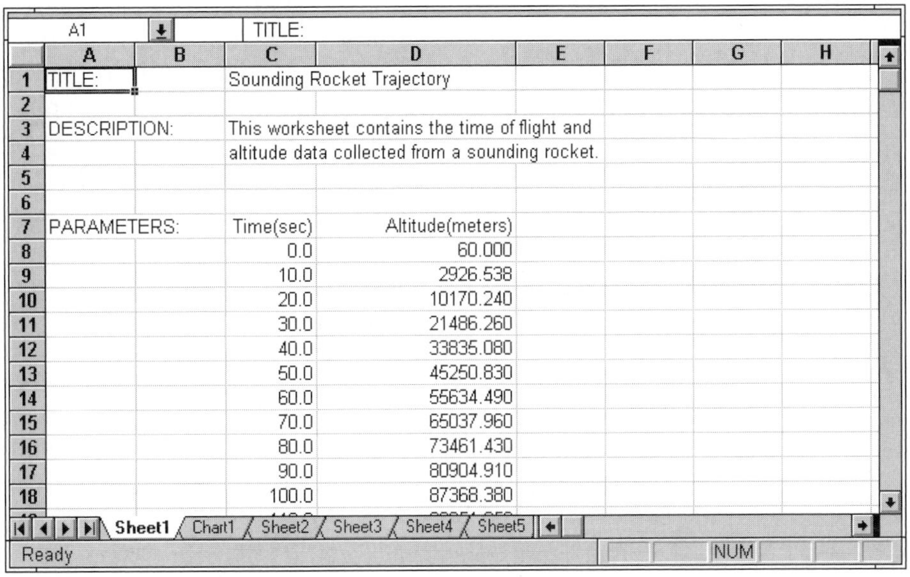

Figure 2-15
TRAJECTORY
Worksheet

	A	B	C	D	E	F	G	H
1	TITLE:		Sounding Rocket Trajectory					
2								
3	DESCRIPTION:		This worksheet contains the time of flight and					
4			altitude data collected from a sounding rocket.					
5								
6								
7	PARAMETERS:		Time(sec)	Altitude(meters)				
8			0.0	60.000				
9			10.0	2926.538				
10			20.0	10170.240				
11			30.0	21486.260				
12			40.0	33835.080				
13			50.0	45250.830				
14			60.0	55634.490				
15			70.0	65037.960				
16			80.0	73461.430				
17			90.0	80904.910				
18			100.0	87368.380				

Figure 2-16
Graph of Altitude Data

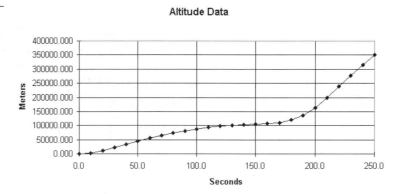

The data points have been collected at ten-second intervals along the actual altitude curve. These data points have been entered into the worksheet so that they are easily available for other analyses. This data is represented in a graph in Figure 2-16. Chapter 4 discusses the graphing capabilities of Excel.

What If

What happens if you change the number of decimal places displayed for altitudes from 3 to 1? Did you actually lose the data? Verify your answer by changing the number of decimal places back to 3.

What happens if you change the format for the altitude to scientific? Why would you want to use a scientific format given that it is not as easy to read as a decimal format?

SUMMARY

This chapter introduced you to the Excel window. You learned how to move the cell pointer within the worksheet; how to enter, edit, and format data; and how to store and retrieve worksheets in data files. The command structure and toolbar buttons were presented along with the Help facility for Excel, which is an online reference manual. By knowing the general categories of commands and by using the extensive Help facility, you should be able to investigate and use new commands on your own after gaining some experience with Excel.

Key Words

active cell	edit line
cascade menu	Excel window
cell	formula bar
cell address	menu bar
cell pointer	mode indicator
clipboard	pull-down menu
contents box	range
control panel	range address
dialog box	sheet tabs

status bar   toolbar
text   value
title bar   workbook

Exercises

The first set of exercises involves modifications to the worksheet examined in this chapter and to the worksheet presented in Chapter 1. Start each problem with the original worksheet.

TRAJECTORY Worksheet

1. Modify the worksheet so that a blank row separates the column titles from the data.

2. Modify the worksheet so that the names of the units (seconds and meters) are in a row below titles TIME and ALTITUDE.

3. Modify the worksheet so that the altitude values are printed in scientific format. Select the scientific format so that all the digits of accuracy of the original data are displayed.

4. Modify the worksheet so that the information is stored in columns C and E instead of columns C and D.

5. Modify the worksheet so that the data values begin in row 10 instead of row 8.

AVERAGE Spreadsheet

6. Change the first data value to 25.46 and observe the change that then occurs in the average value computation.

7. Add an additional blank row before the line with the average value. Thus, there will be two blank rows before the row containing the average value.

8. Add an extra line to the description to indicate that these experimental data values were computed for a chemistry lab experiment that was performed on 1/25/95.

9. Modify the format for the data values to a scientific format.

10. Modify the format for the data values to print only one decimal position of data.

New Worksheets

11. Develop a worksheet that you can use to schedule your week. The schedule should have five columns, with text for each day of the week from Monday through Friday. Include a row for each half hour for the period from 8:00 a.m. to 5:00 p.m.

12. Develop a worksheet that you can use to enter lab data that you will be collecting in your next chemistry laboratory class. Assume that you will be analyzing ten chemical compounds and measuring the pH values and reaction rates.

13. Develop a worksheet that you can use to enter lab data from your next physics laboratory class. Assume that you will be measuring the temperature at which a liquid becomes a gas and that you will be working with five different liquids.

14. Develop a worksheet that you can use to enter lab data from your next botany class. Assume that you will be measuring cell diameters for 20 different specimens.

15. Develop a worksheet that you can use to enter lab data from your next circuits class in which you will be measuring the voltages and currents at five different points in a circuit.

3

Engineering and Scientific Computations

Cable Cars Cable cars or gondolas are often used in ski areas and other mountainous areas to transport people easily and quickly from one location to another. The cable car or gondola operates on a cable that is stretched between two or more supporting towers. The velocity of the cable car depends on its position on the cable relative to a supporting tower. In addition, different equations are used to compute the velocity depending on the distance of the cable car from a cable tower. In this chapter you will use Microsoft Excel to compute the velocity of a cable car.

You are now ready to cover the Excel operations used to perform engineering computations in your worksheets. First you will learn how to include mathematical formulas with the standard operations of addition, subtraction, multiplication, division, and exponentiation. Two types of functions are then discussed—mathematical functions and statistical functions. *Mathematical functions* include trigonometric and logarithmic functions. *Statistical functions* include functions to generate random numbers and to compute averages, variances, and standard deviations. In addition, you will learn about *special functions* that allow you to compare values in the worksheet using logical operations and that allow you to search for information within tables. A final section defines and demonstrates how to add macros to a worksheet. The concept of macros is an advanced topic that you may omit, but macros do provide a powerful way of combining operations into a single keystroke.

Three applications illustrate the formulas and functions presented in this chapter. One worksheet generates uniform random numbers, one computes the velocity for a cable car moving between two towers, and another computes the thermal equilibrium temperatures in a thin metal plate.

3-1 CREATING MATHEMATICAL FORMULAS

In Chapter 2 you learned how to enter data in the cells of a worksheet. Now you will learn how to perform mathematical operations on the data. (In the next section you will learn how to use mathematical functions in formulas.) *Formulas* are entries that perform computations and store the results of the computations in the cells.

Mathematical formulas are similar to algebraic formulas in that symbols and constants are combined to show the computations desired. The formula $Y = A - X + B$ means Y is calculated by subtracting X from A and adding B to the difference. In Excel the values of A, X, and B are entered into specific cells. In the Excel formula the cell addresses for these parameters take the place of the symbols in the algebraic formula. The cell where the formula is entered is the place where the result (the value of Y) is stored. For example, if the values of A, X, and B are stored in cells A1, A2, and A3 respectively, and if Y is represented by cell A4, then the formula entered in cell A4 would be =A1–A2+A3. Any reference to Y in other formulas would use A4, the cell address for Y.

It is essential that all cell references be correct, or the results of the calculations will not be correct. The next discussion covers the different ways to reference the information in cells.

Cell References

A *cell reference* in a formula can be one of three address modes: absolute reference, relative reference, or mixed reference. To indicate an *absolute reference* to cell B4, you would enter $B$4. When you copy a formula containing an absolute reference to another location, the reference still refers to cell B4. A *relative reference* to cell B4 does not use any special characters in the reference. Suppose cell B5 contains a formula with a relative reference to B4, and you copy the formula from cell B5 to cell F21. The relative reference now refers to cell F20, not cell B4, because F20 is in the same relative position to F21 as B4 was to B5. A *mixed reference* contains both types of references. For example, B$4 is a mixed reference that contains a relative column reference and an absolute row reference. If B$4 is contained in a formula, and the formula is copied to another cell, the reference will have a new relative column reference, but the row will always be row 4.

When you are entering or editing a formula, you can press the (F4) function key to change an address mode; each time you press the (F4) function key, the address changes to another mode.

The Formula Bar

The formula bar shown in Figure 3-1 changes as you create formulas. The Name box displays the address of the active cell. It is followed by a drop-down list of range names defined for the workbook, the Cancel box, the Enter box, the Function Wizard, and finally the cell contents box or formula bar. You can use the formula bar buttons to create formulas, edit them, and then confirm or cancel them.

Figure 3-1
Formula Bar

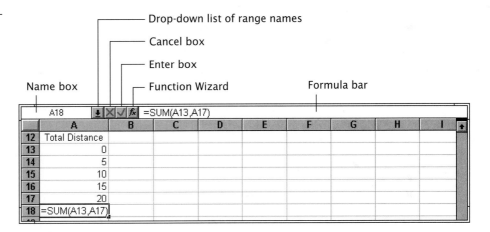

Formula Operations

Formulas allow you to specify the mathematical operations of addition, subtraction, multiplication, division, and exponentiation. For example, if a cell contains the formula =B4+B6, then the contents of cells B4 and B6 are added, and the result is stored in the cell containing the formula. A formula always begins with an equal sign (=). Note that while =B4+B6 is a formula, B4+B6 is text.

Try It

Store values in cells B3 and C3. Then store the formula =B3+C3 in cell D3. Observe the difference between the cell contents box and the data displayed in cell D3. The cell contents box contains the formula entered for cell D3, and the worksheet entry contains the value computed by the formula.

The following rules allow you to create formulas:

Operation	Formula	Example
To add the values in two cells	Use a plus sign (+) between the cell references	=B4+B6
To subtract the values in two cells	Use a minus sign (–) between the cell references	=C2–D4
To multiply the values in two cells	Use an asterisk (*) between the cell references	=C1*C10
To divide the values in two cells	Use a slash (/) between the cell references	=C2/$D$1
To raise the value in a cell to a power	Use a caret (^) between the cell reference and the power	=B4^2

When several mathematical operations are included in a formula, the order of operations is the same as that used in algebra and in most computer languages. Operations in parentheses are performed first, and then the other operations are performed according to the priorities given in Table 3-1. If several operations with the same priority are in the same formula, the operations with the same priority are performed in the order indicated in Table 3-1.

Table 3-1   Arithmetic Operation Priorities

Priority	Operation	Operator	Order
First	Parentheses	( )	inner-most first
Second	Exponentiation	^	right-to-left
Third	Multiplication and division	* /	left-to-right
Fourth	Addition and subtraction	+ –	left-to-right

Try It   Store values in cells D1, D2, D3, and D4. Then use Excel formulas to calculate the following values. Check the values computed using a calculator to be sure you entered the formulas correctly.

♦ $2(D2) - 3.5(D3)$

♦ $D1^2 + D2^3$

♦ $\dfrac{1}{D3 + D4}$

♦ $\dfrac{D1 + D4}{D2 + D3}$

♦ $(D1^2)^3$

Copying Formulas

In many engineering applications, data is stored in rows or columns, and then formulas are applied to the data. For example, suppose you have stored 20 data values in cells A1, A2,..., A20. If you want to square the value of cell A1 and store the result in cell B1, you store the formula =A1^2 in cell B1. If you want to square the value in each of the 20 cells and store the new values in cells B1 through B20, you do not have to enter a new formula in each cell. Instead, you can use the Edit Copy command to copy the formula from cell B1 to the clipboard, and then use the Edit Paste command to copy the formula from the clipboard to cells B2 through B20. Since the formula =A1^2 in cell B1 uses a relative reference, when you copy the formula to cell B2, the formula is automatically modified to refer to the same relative position in column A. Thus, if you copy the formula in cell B1 to cells B2 through B20, the formula is copied and modified so that the formula in B2 is =A2^2, the formula in B3 is =A3^2, and so on. Remember that you should select the range of cells you want to copy before using the Edit Copy command, and then select the destination range before using the Edit Paste command.

Try It

Perform the following steps using formulas and the Edit Copy and Edit Paste commands:

1. Store the values 1 through 25 in cells B1 through B25. (Store only the first value, and then use a formula to compute the rest of the values.)

2. Multiply the value in cell B1 by 4, and store the result in C1. Then multiply each value in cells B2 through B25 by the value in the cell above it, and store the results in cells C2 through C25 (that is, multiply the value in cell B2 by the value in cell B1, and store the result in cell C2, and so on).

3. Subtract the value in cell B1 from each of the values in cells C1 through C25, and store the results in cells F1 through F25. (Be careful.)

Recalculation and Iteration

Each time a value is changed in a cell of a worksheet, all formulas that use that cell are recomputed. Thus, if a number of formulas in a worksheet refer to cell G3, then every time the value in G3 is changed, those formulas are all recalculated. In turn, those recalculations probably cause additional changes. This *automatic recalculation* feature is very powerful because you can make a change and immediately see its effect in the worksheet. If the change did not produce the results that you wanted, you can continue to modify variables until you achieve the desired results.

A formula is normally recalculated only after formulas on which the original formula depends are recalculated. This natural order of recalculation can be changed using the Tools Options command. If you click the Calculation tab in the Options dialog box, the dialog box shown in Figure 3-2 appears.

Figure 3-2
**Calculation Dialog
Box**

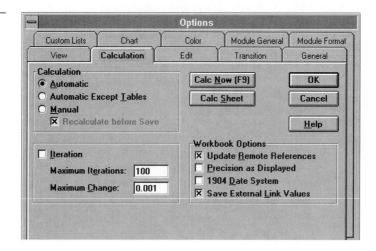

You can change the type of recalculation from automatic to *manual recalculation*, which means changes in cell values will not generate other changes until you press the ⌨F9 function key. This function key causes the worksheet to be recalculated once. In problem solutions that use an iterative algorithm, manual recalculation is preferred; most solutions use automatic recalculation. An example of an iterative solution is given later in this chapter.

Circular References

A *circular reference* occurs when a formula contains a reference that in turn refers to the cell containing the formula. For example, suppose that the following formulas are stored in cells B2 and B3:

Cell B2   =B3+B5

Cell B3   =B2+1

The formula in cell B2 indicates that the values in cells B3 and B5 should be added and the sum stored in cell B2. The formula in cell B3 indicates that the value in cell B2 should be added to 1 and the result stored in cell B3. But a change in the value stored in cell B3 causes a change in cell B2, which causes a change in cell B3, thus creating a circular reference. In general, you do not want to have circular references in your formulas. If a formula contains a circular reference, Excel displays a dialog box indicating that it cannot resolve the circular references. The location of the first cell of the circular reference is displayed on the formula bar. Since a circular reference can be caused by a reference to other cells that contain references back to the first cell, removing a circular reference may require removing one circular reference at a time until there are no additional circular references.

A circular reference is often desired when implementing iterative processes, because you want to repeat steps that refer to the same cell.

Therefore, circular references are usually used in combination with the manual recalculation feature. Later in this chapter an application with an iterative solution uses circular references with manual recalculation of formulas in a row-by-row order.

3-2 USING MATHEMATICAL FUNCTIONS

Engineering problem-solving commonly requires operations such as computing the square root of a value, computing the absolute value of a number, or computing the sine of an angle. Because these types of operations occur so frequently, Excel contains many *functions* that perform common operations.

The name of a function is followed by a set of parentheses that contain the input to the function. This input to the function is also called a *parameter* or an *argument*, and it can be a constant, a range, or even another function reference. Commas are used to separate arguments within a function reference. Spaces are not allowed in a function. The following are some sample function references:

Function with a single argument	=SUM(A1:A10)
Function with two arguments	=SUM(A1:A10,B1:B10)
Nested function	=INT(AVERAGE(A1:A10))

Each type of function is illustrated later in this chapter.

To use a function, you can either type the function in the cell or use the Function Wizard in the formula bar which helps avoid making typographical errors. To use the Function Wizard, first select the cell that will contain the function and type =. When you click the Function Wizard, Step 1 of the Function Wizard dialog box appears as shown in Figure 3-3. If you know the category of the function you want, you can select that category from the Function Category list and then select the function from the Function Name list. If instead you choose the All category, you can select the function name from a list of all functions. After you select a function, you click the Next button, and Step 2 of the dialog box appears. The Step 2 dialog box differs for each function. To obtain instructions on how to complete the dialog box, you can click the Help button. When you are finished, click the Finish button.

Common Functions

Two of the most commonly used functions in engineering are in the Math & Trig function category—the absolute value function and the square root function. Also, since the value of π is so commonly used in engineering computations, Excel contains a function that represents this value; this function does not have any arguments. A function to compute a value of –1, 0, or 1, based on a value's sign, is also useful. The following are common Excel functions. Note that *list* refers to a series of cell addresses or a range.

Figure 3-3
**Function Wizard
Dialog Box**

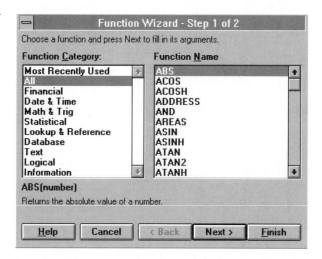

SUM(*list*)	Computes the sum of the values in the list.
AutoSum	Computes the sum of a range. This is such a common function that it has its own button on the formatting toolbar. First select the range to sum and then select the AutoSum button.
RAND( )	Generates a random value between 0 and 1. The same random values are generated in the same order in a work session. The random value changes each time a worksheet is recalculated.
ABS(*x*)	Computes the absolute value of the argument *x*.
SQRT(*x*)	Computes the positive square root of the argument *x*. The argument *x* must be greater than or equal to zero.
PI( )	Gives the value of π as 3.141592654.
SIGN(*x*)	Returns 1 if *x* is positive, 0 if *x* is 0, and –1 if *x* is negative.

Trigonometric Functions

Trigonometric functions and their inverses are available as Excel functions. All the trigonometric functions use radians for angle measurements.

COS(*x*)	Computes the cosine of the argument *x*, which is assumed to be in radians.
COSH(*x*)	Computes the hyperbolic cosine of the argument *x*.
SIN(*x*)	Computes the sine of the argument *x*, which is assumed to be in radians.
SINH(*x*)	Computes the hyperbolic sine of the argument *x*.
TAN(*x*)	Computes the tangent of the argument *x*, which is assumed to be in radians.

TANH(*x*)	Computes the hyperbolic tangent of the argument *x*.
ACOS(*x*)	Computes the arc cosine, or inverse cosine, of the argument *x*, which must be between –1 and 1. The angle returned is in radians and will be between 0 and π.
ACOSH(*x*)	Computes the inverse hyperbolic cosine of *x*.
ASIN(*x*)	Computes the arc sine, or inverse sine, of the argument *x*, which must be between –1 and 1. The angle returned is in radians and will be between –π/2 and π/2.
ASINH(*x*)	Computes the inverse hyperbolic sine of *x*.
ATAN(*x*)	Computes the arc tangent or inverse tangent of the argument *x*. The angle returned is in radians and will be between –π/2 and π/2.
ATAN2(*x*,*y*)	Computes the arc tangent or inverse tangent of the value *y*/*x*. The angle returned is in radians and, depending on the signs of *x* and *y*, will be between –π and π.
ATANH(*x*)	Computes the inverse hyperbolic tangent of *x*.
RADIANS(*x*)	Converts degrees to radians. The argument *x* is in degrees, and the angle returned is in radians. (This function is very useful because many of the trigonometric functions assume that the arguments are in radians.)
DEGREES(*x*)	Converts radians to degrees. The argument *x* is in radians, and the angle returned is in degrees.

Exponential and Logarithmic Functions

Exponential and logarithmic functions are also available as Excel functions. Recall that since the logarithm of a negative number does not exist, the logarithmic functions LN and LOG can only be used with positive arguments.

EXP(*x*)	Computes the value of *e* raised to the *x*th power.
LN(*x*)	Computes the natural logarithm of *x* (which is the logarithm of *x* to the base *e*). The argument *x* must be greater than 0.
LOG(*x*)	Computes the common logarithm of *x* to the base of 10. The argument *x* must be greater than 0.

When using one function as an argument for another function, be sure to enclose the argument of each function in its own set of parentheses. This *nesting* of functions is also called *composition* of functions. For example, the following function reference guarantees that the argument is positive:

=LOG(ABS(C2))

Rounding and Truncating Functions

In engineering applications you often need to perform operations such as rounding values to the nearest integer or truncating values by dropping any decimal portion. These Excel functions allow you not only to perform truncating and rounding but also to determine the quotient and remainder values in division.

INT(x)	Returns the integer part of the argument x. It does not perform any rounding.
MOD(x,y)	Returns the remainder of the division of x by y. The value of y cannot be 0, and the sign of the value returned is the same as the sign of x.
ROUND(x,n)	Rounds x to the position specified by 10^{-n}.

Suppose that you want to round a group of numbers to the nearest hundred. Thus 256 should round to 300, and –312 should round to –300. You can perform this operation on the value in cell B5 using the following function:

$$=ROUND(B5,-2)$$

Note that you need to store the rounded values in new cells to avoid circular references.

Try It

Perform the following steps using functions for as much of the work as possible:

1. Generate a column of values that contain angles in degrees starting with 0 degrees and incrementing by 10 degrees, with the last cell containing 350 degrees. Start the column at cell C1.

2. Convert the column of angles in degrees to angles in radians, and store the corresponding column starting in cell E1.

3. Compute the tangent of each angle in column E, and store the result in column F, starting in cell F1.

4. Compute the inverse tangent of each value in column F by using ATAN, and store the results in column G. Explain any differences between the angles in column G and those in column E.

5. Round the angles in radians in column E to the nearest thousandth, and store the results in column H.

3-3   INCLUDING STATISTICAL COMPUTATIONS

A number of computations that are considered to be statistical computations are very useful in solving engineering problems. These computations include computing the average of the values in a list, computing the maximum value in a list, and computing the minimum value in a list. Several of these computations have been incorporated into Excel's Statistical function category.

Statistical Functions

AVERAGE(*list*)	Computes the average of all values in the list.
COUNT(*list*)	Counts the number of cells in the list.
MAX(*list*)	Returns the maximum value from the values in the list.
MIN(*list*)	Returns the minimum value from the values in the list.
STDEV(*list*)	Computes the sample standard deviation from the values in the list.
STDEVP(*list*)	Computes the population standard deviation from the values in the list.
VAR(*list*)	Computes the sample variance from the values in the list.
VARP(*list*)	Computes the population variance from the values in the list.

Most of these statistical functions should represent familiar computations. However, since you may not have worked with standard deviations and variances, definitions and examples of these terms will be given in this section and in the next section.

The AVERAGE function computes the *average* (also called the *mean*) value of a list of values. Thus, if you enter the function reference =AVERAGE(1,2,3) in a cell, the value is computed to be the sum of the values (1+2+3) divided by 3, or a mean value of 2. The population variance of the same list of values is the average squared deviation of the values from the mean value. Since the deviation of a value from the mean is the same as the difference between the value and the mean, the variance can also be defined as the average of the squared differences from the mean. Thus, if you enter the function reference VARP(1,2,3) in a cell, the value is computed to be 0.666667, or 2/3. You can compute this value by hand with the following equation (recall that the mean of the values is 2):

$$
\begin{aligned}
\text{variance} &= \frac{(1-2)^2 + (2-2)^2 + (3-2)^2}{3} \\
&= \frac{(1+0+1)}{3} \\
&= \frac{2}{3}
\end{aligned}
$$

The *standard deviation* is the square root of the variance. Thus, if you enter the function reference STDEVP(1,2,3), the value is computed to be the square root of 0.666667, or 0.816497. This example shows you how to compute the variance and standard deviation of a list of values. The next section discusses random numbers, and you will see how the variance and standard deviation are used in analyzing random signals or noise. (We use the population variance and standard deviation in our examples. A sample variance is computed by dividing the sum of the squared differences from the mean by n–1, where n is the number of points. Sample variances and standard deviations are used in some engineering applications.)

Range Names

Since the argument of most statistical functions is a list, it is sometimes helpful to use a range reference instead of listing the range; that is, you can substitute a *range name* for an address. For example, suppose you have created a worksheet that contains a set of time values and temperature values. Assume that the time values are contained in the range A14:A33 and the temperature values are contained in the range C14:C33. To compute the average temperature, you could use the following function reference:

$$=\text{AVERAGE(C14:C33)}$$

However, if you had created a range name called TEMPS for the range C14:C33, you could use the function reference

$$=\text{AVERAGE(TEMPS)}$$

Not only is it easier to remember what you are computing with the range reference, but you can also avoid mistakes because you do not need to enter the range address each time you want to use the temperature data.

To create a range name, first select the range. You can then either type the range name in the name box or choose the Insert Name command. Type the name in the Define text box, and then select the Add button. The range names are saved when you save a worksheet. If a reference to a range name is used in a formula, a dollar sign is often used in front of the range name to specify that it is an absolute reference. To display a list of all range names in the worksheet, click the drop-down list on the edit line.

Try It

Load the worksheet TRAJECTORY from the data files available with this module. Define a new range name for the column of altitude data. Now use cells J16 through J21 to compute the following values (using the range name in the function references):

- Average of the altitude values
- Number of the altitude values
- Maximum of the altitude values
- Minimum of the altitude values
- Standard deviation of the altitude values
- Variance of the altitude values

Application 1   **RANDOM NUMBER GENERATION**

Electrical/Computer Engineering

As we saw in the last section, the RAND function generates numbers that are equally likely to occur between 0.0 and 1.0. However, random numbers that range between values other than 0.0 and 1.0 are often needed.

Bounds for Random Numbers

You can generate a random number that is evenly, or uniformly, distributed between a lower bound and an upper bound by modifying the number generated by the RAND function. First multiply (or scale) the number generated by the RAND function by the width of the distribution of the desired random number. This width is computed by subtracting the lower bound from the upper bound. Then add the scaled value to the lower limit to adjust the new value to the proper range. For example, suppose that you want to generate values between 2 and 5. Use the RAND function to generate a number between 0 and 1, and then multiply it by 3, which is the difference between the upper and lower bounds (5–2). Add the lower bound (2), giving a resulting value that is equally likely to be any value between 2 and 5. Thus, a corresponding formula to perform these steps is =3*RAND()+2.

Worksheet for Generating Uniform Random Numbers

You are now ready to develop a worksheet as outlined in Figure 3-4 to generate 100 uniform random numbers between specified lower and upper bounds.

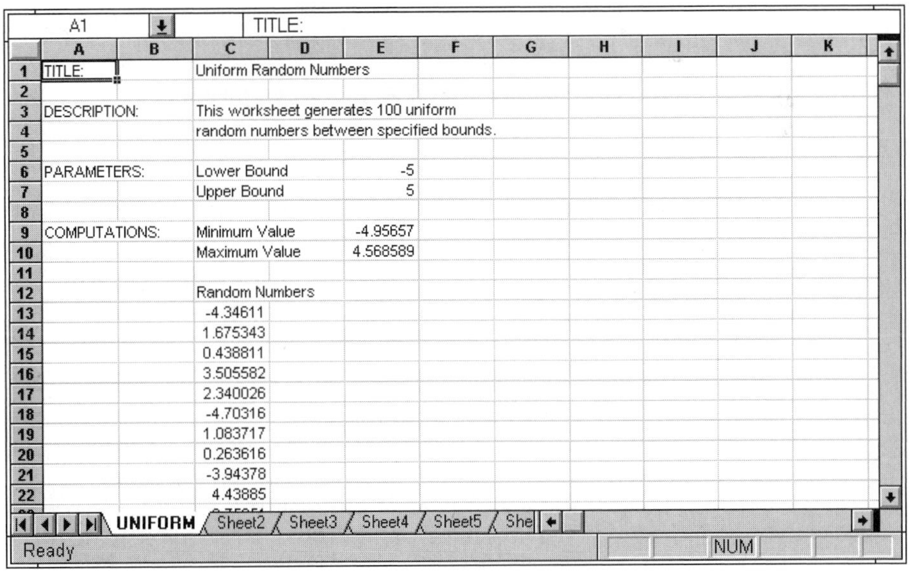

Figure 3-4
UNIFORM Worksheet

	A	B	C	D	E	F	G	H	I	J	K
1	TITLE:		Uniform Random Numbers								
2											
3	DESCRIPTION:		This worksheet generates 100 uniform								
4			random numbers between specified bounds.								
5											
6	PARAMETERS:		Lower Bound		-5						
7			Upper Bound		5						
8											
9	COMPUTATIONS:		Minimum Value		-4.95657						
10			Maximum Value		4.568589						
11											
12			Random Numbers								
13			-4.34611								
14			1.675343								
15			0.438811								
16			3.505582								
17			2.340026								
18			-4.70316								
19			1.083717								
20			0.263616								
21			-3.94378								
22			4.43885								

UNIFORM / Sheet2 / Sheet3 / Sheet4 / Sheet5 / She

Ready   NUM

1. Problem Statement

Generate 100 uniform random numbers.

2. Input/Output Description

Begin the worksheet using a solution template that includes a title, a description, and the parameters of the worksheet, as shown in Figure 3-4. To use this worksheet, you enter the lower and upper bounds for the random numbers. The worksheet should compute the random numbers. It should also determine the minimum and maximum values from the random numbers.

3. Hand Example

Using the steps presented earlier in this section, transform a random number (call it the old RN) from a uniform distribution between 0 and 1 to a random number (call it the new RN) from a uniform distribution between an upper and lower bound with the following formula:

$$\text{new RN} = (\text{upper} - \text{lower}) * (\text{old RN}) + \text{lower}$$

Thus, if the old RN is 0.234, and you want to transform this to a value between –5 and 5, use the following computation:

$$\text{new RN} = (5 - (-5)) * (0.234) + (-5) = -2.66$$

4. Algorithm Development

The worksheet now contains everything except the formulas to compute random numbers. The value of the lower bound should be stored in cell E6, and the value of the upper bound should be stored in cell E7. Select cell E6 and type LOWER in the name box to name the first location. Repeat these steps to name cell E6 UPPER. The random numbers are generated using the steps outlined in the hand example:

| C13 | =RAND()*(UPPER–LOWER)+LOWER |
| C14:C112 | Copied from C13 |

To name the random numbers NOISE, select the random numbers, and type NOISE in the name box. Then use this reference in computing the statistics:

| E9 | =MIN(NOISE) |
| E10 | =MAX(NOISE) |

5. Testing

Figure 3-4 shows the worksheet with a typical set of random numbers in it. Each time the worksheet is recalculated, all references to RAND generate new values. Therefore, if you change either the lower or upper bound, the values are immediately recalculated using the new bound. If you replace the lower or upper bound with the same value, the values are still recalculated, and the random numbers will change. You can also use the (F9) function key to recalculate the values. It is interesting to change one of the parameters and observe the corresponding changes in these statistics. This worksheet is saved as UNIFORM.XLS on the data diskette available with this module.

What If

Load the UNIFORM worksheet and use the (F9) function key to recalculate it ten times. What was the minimum value for the group of ten runs?

3-4   USING SPECIAL FUNCTIONS

In addition to the mathematical and statistical functions already presented in this chapter, Excel also contains several special-purpose functions. The IF function and the LOOKUP functions are the focus of this section because they are very useful in solving engineering problems. The DATE and TIME functions are also included.

IF Function

The IF function is part of the Logical function category and is particularly useful because it allows you to evaluate conditions using cell values. The function and its arguments are

IF(*logical test, value_if_true, value_if_false*)

To compute the function value, *logical test* is evaluated to see if it is true or false. If it is true, *value_if_true* is stored in the cell containing the function reference. If the condition is false, *value_if_false* is stored in the cell containing the function reference. To illustrate, assume that cell B3 contains the following function:

=IF(B1>0,1,–1)

If the value in B1 is greater than zero, then the value 1 is stored in cell B3. If the value in B1 is less than or equal to zero, then the value –1 is stored in cell B3.

The conditions tested in the IF function use the following logical operators:

=	Equal to
<	Less than
>	Greater than
< =	Less than or equal to
> =	Greater than or equal to
< >	Not equal to

The logical operator is used to compare two values, and the resulting expression is always evaluated to be either true or false. If the expression is true, the first value after the condition is stored in the cell; if the expression is false, the last value in the function reference is stored in the cell.

Two conditions can be combined with logical functions AND or OR, as in OR(B2>B5,B2>F5). When you join two conditions with AND, the combined condition is true only if both individual conditions are true. When you join two conditions with OR, the combined condition is true if either or both individual conditions are true. Conditions can also be preceded by NOT, as in NOT(A=B). When logical operators are combined with AND, OR, and NOT, the logical expressions are evaluated first, followed by NOT, and finally by AND and OR.

The IF function is very useful in developing a worksheet that asks a question and then determines the correct value to compute based on the answer to that question. In the next section you will compute the velocity of a cable car. The velocity is computed with one equation if the cable car is within 30 feet of a tower; otherwise, a different equation is used. The IF function allows you to choose the correct equation.

LOOKUP Functions

The LOOKUP functions can also be very useful in solving engineering problems. These functions are part of the Lookup & Reference function category and allow you to specify a value that you want to look up in a range of values that are referred to as a table. If you want to match a value in a column of the table and then use information from the corresponding row, the table is a vertical table, and you should use the VLOOKUP function. If you match a value in a row of the table and then use information from the corresponding column, the table is a horizontal table, and you should use the HLOOKUP function.

The VLOOKUP function has three parameters: the *lookup_value* for which you are looking in the first column, the *table_array* that will be searched, and the *col_index_num* from which the matching value will be returned.

VLOOKUP (*lookup_value, table_array, col_index_num*)

The first column of data in the *table_array* must contain information in ascending order. When the VLOOKUP function is referenced, the value of *lookup_value* is compared to the values in the first column of the table. If *lookup_value* is less than the smallest value in *col_index_num*, an #N/A!

error occurs. Otherwise, VLOOKUP identifies the row corresponding to the largest value in the first column that is less than or equal to *lookup_value*. VLOOKUP then returns the value located in that row using the column identified by *col_index_num*. VLOOKUP returns a #VALUE! error if the *col_index_num* is less than 1 and a #REF! error if *col_index_num* exceeds the number of columns in *table_array*. The HLOOKUP function works in a similar manner, working with the values in a row of an array for a match and returning the value of the indicated cell in the corresponding column.

To illustrate the VLOOKUP table, use the following portion of a steam table[1] that is commonly used in chemical engineering calculations. A steam table contains the thermodynamic properties of steam and includes values for entropy, internal energy, enthalpy, and other properties of steam over a range of pressures and temperatures. This information is useful in the analysis and design of reciprocating steam engines, turbines, and refrigerators. The following table contains temperature, volume, enthalpy, and entropy for superheated steam at a pressure of 250 lb/in^2:

Temperature	Volume	Enthalpy	Entropy
420	1.9077	1214.2	1.5414
440	1.9717	1227.3	1.5560
460	2.033	1239.7	1.5697
480	2.093	1251.7	1.5826
500	2.151	1263.4	1.5949
520	2.208	1274.8	1.6067

If this table were stored in a range named TABLE, and if you wanted to select the volume that goes with a specific temperature stored in cell A6, you would use the following function reference:

$$=VLOOKUP(A6,TABLE,2)$$

The volume column is referenced with *col_index_num* of 2. Thus, when the value in cell A6 is 500, the function value is 2.151. Also, when the value in cell A6 is 503, the function value is still 2.151. If you wanted to select enthalpy values, the reference would be:

$$=VLOOKUP(A6,TABLE,3)$$

When the value in cell A6 is 580, the enthalpy value is 1274.8.

Excel offers a number of functions in addition to the ones presented in this chapter—over 200 in fact. You can use online Help to find out more about all the Excel functions.

Try It

Assume that cell B20 contains a reference to an IF function. Give the complete IF function that performs the following:

---

[1] Sears, F. W. *An Introduction to Thermodynamics, the Kinetic Theory of Gases, and Statistical Mechanics*, Second Edition. Reading, Mass.: Addison-Wesley, 1964.

◆ If the value in cell B2 is greater than the value in cell B3, then cell B20 should contain zero; otherwise, cell B20 should contain 10.

◆ If the value in cell B3 is not equal to 100, then cell B20 should contain the sum of A10 and B10; otherwise, cell B20 should contain the difference between A10 and B10.

◆ If the value in cell B1 is greater than B2, or if the value in cell F6 is not equal to that in cell G6, then cell B20 should contain C10; otherwise, cell B20 should contain the square root of C10.

DATE and TIME Functions

The DATE function, which is in the Date & Time function category (along with the TIME function), has three arguments that represent year, month, and day, as shown in the following reference:

$$=DATE(95,12,31)$$

Excel converts the date into a date number using the 1900 date system: The number is an integer between 1 and 65380, representing the days between 1 January 1900 and 31 December 2078. The date can be displayed in a variety of forms. First select the cell, and then use the Format Cells command. Next select how you would like to have the date displayed from the choices in the Number tab. Some of the choices are

12/31/95

31-Dec-95

31-Dec

Dec-95

The TIME function also has three arguments that represent the hour (in 24-hour form), minutes, and seconds, as shown in the following reference:

$$=TIME(13,10,25)$$

The time can be displayed in a variety of forms. First select the cell, and then use the Format Cell command and select one of the forms displayed in the Number tab. Some of the choices are

1:10 PM

1:10:25 PM

13:10

13:10:25

10:25

These two functions allow you to add date and time information to your reports and worksheets.

Application 2	**CABLE CAR VELOCITY**

Mechanical Engineering

In this application you compute the velocity of a cable car as it moves along a 500-foot cable stretched between two towers, as shown in Figure 3-5. You want to prepare a table of distance and velocity values using equations that relate the distance of the cable car from a tower to its corresponding velocity.

Figure 3-5
Cable Car Towers

Tower 1   Tower 2

Cable Car Velocity Equations

The velocity of the cable car depends on its position on the cable; assume that you have been given equations that approximate the velocity using the distance of the cable car from a tower. When the cable car is within 30 feet of a tower, its velocity is approximated by the following equation:

$$\text{velocity} = 2.425 + 0.00175\,d^2\,\text{ft/sec}$$

where d is the distance in feet from the cable car to the nearest tower. If the cable car is not within 30 feet of a tower, its velocity is approximated by this equation:

$$\text{velocity} = 0.625 + 0.12\,d - 0.00025\,d^2\,\text{ft/sec}$$

You want to prepare a table starting with the cable car at the first tower and moving to the end tower in increments of 5 feet.

Worksheet for Generating Cable Car Velocity Values

In the worksheet you will increment the distance of the cable car by 5 feet for each new computation. You also want to include the number of the nearest tower (1 = first, 2 = end), the distance from the nearest tower, and the velocity of the cable car at that position. (Note that it is the distance from the nearest tower, not the total distance, that is used in the velocity equations.)

 1. Problem Statement

Generate a table showing the velocity of a cable car as it moves between two towers.

2. Input/Output Description

Begin the worksheet using a solution template that includes a title, a description, and the parameters of the worksheet, as shown in Figure 3-6.

Figure 3-6
CABLE worksheet

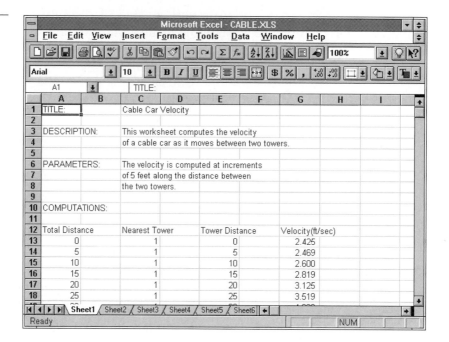

 3. Hand Example

The total distance traveled by the cable car from the first tower begins at zero and increases in increments of 5 feet until it reaches the second tower. Therefore, the cable car is closer to Tower 1 when the total distance of the cable car from Tower 1 is 0 to 250 feet. When the total distance is greater than 250 feet, the cable car is closer to Tower 2. Note that when the total distance is 250 feet, you could choose either Tower 1 or Tower 2.

Furthermore, the distance to the nearest tower is the total distance when the total distance is less than or equal to 250 feet. The distance to the nearest tower is 500 feet minus the total distance when the total distance is greater than 250 feet. A few examples illustrate these computations:

Total Distance	Distance to Nearest Tower
200	200
300	500 – 300 = 200
450	500 – 450 = 50

4. Algorithm Development

Now you need to add the formulas to compute the information in the four columns. The total distance values begin with 0 and increase in increments of 5, ending with 500. Thus the table contains a total of 101 values. You can compute the total distance values with the following formulas:

A13	0
A14	=A13+5
A15:A113	Copied from A14

To determine which tower is nearest, use the IF function. If the distance is less than 250 feet, the cable car is closer to Tower 1; otherwise, it is closer to Tower 2:

C13	=IF(A13<250,1,2)
C14:C113	Copied from C13

To compute the distance to the nearest tower, you need another IF function. If the distance is less than or equal to 250, the distance to the nearest tower is just the total distance. Otherwise, the distance to the nearest tower is 500 feet minus the total distance:

E13	=IF(A13<=250,A13,500−A13)
E14:E113	Copied from E13

Finally, compute the velocity using still another IF function. The distance to the nearest tower determines which equation you use.

G13	=IF(E13<=30,2.425+0.00175*E13^2, 0.625+0.12*E13−0.00025*E13^2)
G14:G113	Copied from G13

5. Testing

The final form of the worksheet (with column G formatted with three decimal places) is saved as CABLE.XLS on the diskette available with this module.

What If

Using the CABLE worksheet, modify the distance computations so that the increment is by 2.5 feet instead of 5 feet. Are the rest of the computations correct? You can compare some of the values computed with ones that have the same total distance from the original worksheet.

Application 3	**TEMPERATURE DISTRIBUTION**

Mechanical Engineering

This application considers the temperature distribution in a thin metal plate as it reaches a point of thermal equilibrium. The plate is constructed so that each edge is isothermal (maintained at a constant temperature). The temperature of an interior point on the plate is a function of the temperature of the surrounding material. If you consider the plate to be similar to a grid, then the cells of a worksheet could be used to store the temperatures of the corresponding points on the plate. Figure 3-7 contains a grid that could be used to store the temperatures of a plate that is being analyzed with five temperature measurements along the sides and ten temperature measurements along the top and bottom.

Figure 3-7
Temperature Grid for a Thin Metal Plate

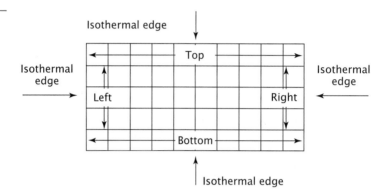

Thermal Equilibrium

The isothermal temperatures at the top, bottom, left, and right of the grid are fixed parameters. The interior points are initially set to some arbitrary temperature, usually zero. The new temperature of each interior point is calculated as the average of its four surrounding points, as shown in the following equation:

$$T_0 = \frac{(T_1 + T_2 + T_3 + T_4)}{4}$$

	T_1	
T_4	T_0	T_2
	T_3	

Each iteration updates all the internal points. These iterations should continue until the interior points quit changing.

It should be clear that you need to turn off the automatic recalculation feature to implement this iterative process with Excel. Otherwise, the formula for a point in the interior updates its point, and that in turn affects the neighboring points, which affects the original interior point again.

Worksheet for Computing Temperature Distribution

You are now ready to develop a worksheet to implement this iterative process.

1. Problem Statement

Determine the equilibrium values for a metal plate.

2. Input/Output Description

Begin the worksheet using a solution template that includes a title, a description, and the parameters of the worksheet, as illustrated in Figure 3-8. The isothermal temperature values go in cells D8 through D11. The grid for storing the temperature values for a plate with four rows and four columns uses columns C, D, E, and F and rows 16 through 19. You also need an additional parameter that is initially set to zero. If you set this parameter to 1, another parameter will begin counting the number of iterations performed.

Figure 3-8
PLATE Worksheet with Initial Temperatures

	Microsoft Excel - PLATE.XLS

File   Edit   View   Insert   Format   Tools   Data   Window   Help

A1 | TITLE:

	A	B	C	D	E	F	G	H	I	J
1	TITLE:		Temperature Distribution							
2										
3	DESCRIPTION:		This worksheet initializes the							
4			temperatures in a metal plate and							
5			determines the equilibrium temperature.							
6										
7	PARAMETERS:		Isothermal Temperatures:							
8			Top	100						
9			Bottom	200						
10			Left	100						
11			Right	200						
12										
13			Initial	0	Iteration	0				
14										
15	COMPUTATIONS:		4X4 Grid							
16			100	100	100	100				
17			100	0	0	200				
18			100	0	0	200				
19			200	200	200	200				

Sheet1 / Sheet2 / Sheet3 / Sheet4 / Sheet5 / Sheet6

Ready   Calculate   NUM

Since you want to control the recalculation feature in this worksheet, choose Tools Option and select the Manual Calculation button on the Calculation tab.

3. Hand Example

Assume that the plate has isothermal edge temperatures such that the top and left sides are 100.0 and the bottom and right sides are 200.0. Also assume that the internal points are initialized to zero. Following is the initial temperature grid:

100.0	100.0	100.0	100.0
100.0	0.0	0.0	200.0
100.0	0.0	0.0	200.0
200.0	200.0	200.0	200.0

Assume that the computations are in a row-by-row order so that after the first iteration the temperature grid is as follows:

100.0	100.0	100.0	100.0
100.0	50.0	87.5	200.0
100.0	87.5	143.75	200.0
200.0	200.0	200.0	200.0

You can continue performing iterations until the interior points quit changing, which is the thermal equilibrium point. You can also stop the iterations when the interior temperatures are changing by a small value.

 ## 4. Algorithm Development

The worksheet now contains everything except the formulas to initialize the edge temperatures, compute the interior temperature values, and update the iteration count. The formulas for initializing the edge temperatures are

C16 $D$8	D16 $D$8	F16 $D$8
C17 $D$10	D19 $D$9	F17 $D$11
C18 $D$10	E16 $D$8	F18 $D$11
C19 $D$9	E19 $D$9	F19 $D$9

In developing the formula for the internal temperatures, you should provide a way to reinitialize the internal temperatures to zero. Therefore, use the initialization parameter stored in D13. Whenever this value is zero, you set the internal temperatures to zero and reset the iteration count to zero. When D13 is not zero, a recalculation with the (F9) (CALC) function key increments the iteration count and performs the computations to update the interior temperatures. The corresponding formulas are

F13   =IF($D$13=0,0,$F$13+1)

D17   =IF($D$13=0,0,(D16+E17+D18+C17)/4)

D18, E17, E18 Copied from D17

5. Testing

The worksheet is saved as PLATE.XLS. When testing this worksheet, initialize the process by storing a value of zero in cell D13. Figure 3-8 contains the worksheet with the parameters initialized but no iterations performed.

Change the initialization parameter to 1, and use the (F9) (CALC) function key to recompute the worksheet. Figure 3-9 contains the worksheet after one iteration; note that these values are the same values computed in the hand example.

Figure 3-9
PLATE Worksheet after One Iteration

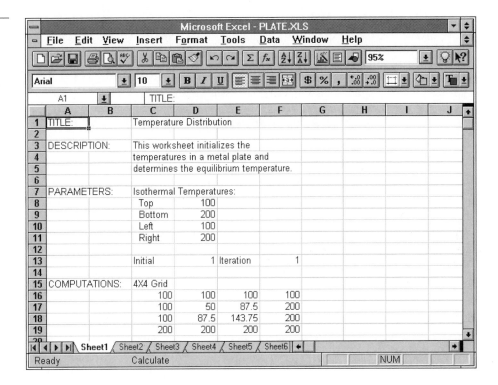

You can continue recomputing the worksheet until the interior temperatures reach equilibrium. For this set of parameters, thermal equilibrium occurred after 12 iterations, as shown in Figure 3-10.

What If How many iterations are necessary to reach equilibrium if all edge temperatures are set to 200? (Use the PLATE worksheet from the diskette available with this module to answer this question.)

Figure 3-10
PLATE Worksheet at Equilibrium

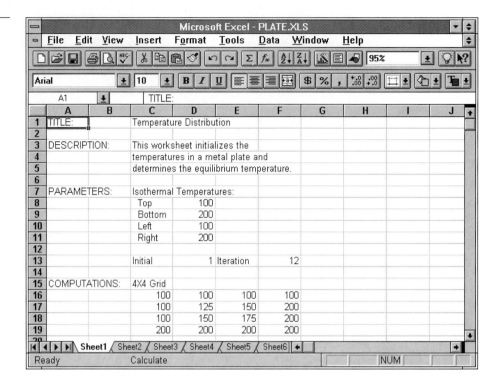

3-5   CREATING MACRO COMMANDS

A *macro command* (or just *macro*) is a series of commands that you enter into a cell (or cells) in a worksheet. You can then perform the commands simply by using a menu or shortcut key, thus automating steps that you frequently use in a worksheet. The macro is saved as part of the worksheet and is available whenever the worksheet is loaded.

Before outlining the details of creating and using a macro command, a simple example is useful. Suppose you would like each column of a worksheet you are developing to be wide enough to fit its widest entry. The command to do this takes several steps. You begin by selecting the cells used in the worksheet. You then choose Format Column AutoFit. Several motions are required, and if you perform this step several times as you continue to develop your worksheet, then this operation is a good candidate for a macro command. When you create a macro to store the steps for setting each column width to fit its widest entry, you only need to invoke the macro using the Macro menu.

Macros are created by recording the steps that you take to perform a task. To start recording, you choose Tools Record Macro. You then choose Record New Macro. The Record New Macro dialog box appears. Excel automatically names your macro, but giving it a more descriptive name than MACRO1 is a good idea. Once you've given your macro a name, you enter a description of it, so that you'll remember what it does, or so that someone else can use it. Click OK to start recording your macro. Notice that the mode indicator displays Recording.

To record your macro, perform the operations that you want saved in the macro, and the macro recorder captures your keystrokes. When you are finished, click the Stop button that has appeared on the upper-right corner of your worksheet.

To demonstrate how to create the macro that sets each column width in the range A1:G20 to fit its largest entry, you begin by choosing Tools Record Macro, and then Record New Macro. You next name the macro FIT_WIDTH.

Provide documentation under Description that describes the macro as working on columns A1:G20 and ensuring that the columns are large enough for the longest entry in each. Then click OK. Select cells A1:G20. Then select the Format Column Autofit option. The last step is to click the Stop button.

To run your macro, you choose Tools Macro menu and select the name FIT_WIDTH. Then choose the Run command button. To learn more about creating macros, see the *Microsoft Excel Visual Basic User's Guide* for details.

SUMMARY

You can perform engineering computations in Excel using formulas and functions. You use formulas when the operations that you perform use the standard operations of addition, subtraction, multiplication, division, and exponentiation. You use functions for other computations, ranging from trigonometric functions to logarithmic and statistical functions. Formulas can include function references, and function references can include formulas. Thus the combination provides a very powerful tool for performing engineering computations. In this chapter you developed worksheets to generate uniform random numbers, to compute cable car velocities, and to determine the thermal equilibrium of a thin metal plate. Finally, you learned how to create macros to save keystrokes.

Key Words

absolute reference	mathematical function
argument	mean
automatic recalculation	mixed reference
average	nesting
cell reference	parameter
circular reference	range name
composition	relative reference
formula	special function
function	standard deviation
macro command (macro)	statistical function
manual recalculation	

Exercises

The first set of exercises involves modifications to the worksheets generated in this chapter and the previous chapter. Start each problem with the original worksheet.

CABLE Worksheet

1. Modify the worksheet so that the distance increment is 10 feet, unless the cable car is less than 20 feet from a tower, in which case the distance increment is 5 feet.

2. Modify the worksheet so that it contains the maximum velocity, the minimum velocity, and the corresponding total distances.

PLATE Worksheet

3. Modify the worksheet so that the computations are performed in column-by-column order. Does this change the number of iterations necessary to reach thermal equilibrium for the hand example?

4. Modify the worksheet so that the grid contains six rows and eight columns.

TRAJECTORY Worksheet

5. Modify the worksheet so that it computes a third column for the altitude in feet.

6. Modify the worksheet so that it determines the two consecutive altitudes with the maximum change between them.

New Worksheets

7. One problem in timber management is to determine how much of an area to leave uncut so that the harvested area is reforested in a certain period of time. It is assumed that reforestation takes place at a known rate per year, depending on climate and soil conditions. A reforestation equation expresses this growth as a function of the amount of timber standing and the reforestation rate. For example, if 100 acres are left standing after harvesting, and the reforestation rate is 0.05, then 100 + 0.05(100), or 105 acres, are forested at the end of the first year. At the end of the second year, the number of acres forested is 105 + 0.05(105), or 110.25 acres. Develop a worksheet named TIMBER that uses parameters that specify the number of acres, the number of acres that are uncut, and the reforestation rate. The worksheet should generate a report that tabulates for 20 years the number of acres reforested and the total number of acres forested at the end of each year.

8. Modify the spreadsheet developed in problem 7 so that the number of years for which the report is generated is a parameter that is specified before the computations section of the spreadsheet.

9. A rocket is being designed to test a retrorocket intended to permit softer landings. The designers have derived the following equations that they believe will predict the performance of the test rocket, where t represents the elapsed time in seconds:

$$\text{acceleration} = 4.25 - 0.015t^2$$

$$\text{velocity} = 4.25t - 0.005t^3$$

$$\text{height} = 90 + 2.125t^2 - 0.00125t^4$$

The height equation gives the height above ground level at time t. Thus the first term (90) is the height in feet above ground level of the nose of the rocket at launch. To check the predicted performance, the rocket will be "test flown" with the computer, using the derived equations. Develop a worksheet to print the time, height, velocity, and acceleration for the rocket from time of zero seconds through 50 seconds, in increments of 1 second.

10. Modify the spreadsheet developed in problem 9 so that the time increment is 1 second from a time of zero seconds through 40 seconds and is 0.5 second from 40 seconds through 50 seconds.

11. A biologist has just determined the characteristics of a new bacterium. This bacterium has a constant growth rate. If 10 cells are present with a growth factor of 0.1, the next generation will have $10 + 10(0.1) = 11$ cells. Develop a worksheet that computes and prints a report of the number of cells and the percent area of petri dish covered for each of five generations. (Ten cells occupy 1 square millimeter.) The input parameters needed to compute the report are the number of cells initially, the petri dish diameter, and the growth rate. The number of cells should always be an integer.

12. Modify the spreadsheet developed in problem 11 so that the growth factor is a parameter that is entered in the worksheet.

13. Develop a spreadsheet that will compute the values of the following polynomial for a specified set of time values:

$$f(t) = 2.5 - t^2 + 3.6t^3 - 0.5t^4$$

The parameters of the worksheet should include the starting value of t in seconds and the increment between time values in seconds. A total of 100 time values should be used. The spreadsheet should include in the computations section a table of the corresponding time values and polynomial values.

14. Insert a section at the beginning of the computations section of the spreadsheet developed in problem 13 that includes the maximum and minimum polynomial values computed for the polynomial.

15. Add a third column to the values computed in the spreadsheet developed in problem 13. Fill this column with random values between –0.5 and 0.5. Then, add a fourth column that contains the sum of the corresponding polynomial values and the random number values.

16. Insert a section at the beginning of the computations section of the spreadsheet developed in problem 15 that includes the average value and the standard deviation from the third column of information which is random numbers.

4 Engineering Graphs

Microprocessors

Adigital filter extracts information from a digital signal. For real-time applications, digital filters are often implemented in microprocessors or in VLSI (very large scale integration) chips on printed circuit boards, as shown here. For other applications, the signals are collected and then filtered at a later time. Microsoft Excel provides a convenient tool for storing the digital signals, performing the filtering operations, and plotting the various signals to see the effects of the filters.

INTRODUCTION

It is hard to imagine performing engineering or scientific experiments without the visual information incorporated in graphs. Microsoft Excel has the capability to generate many different types of graphs or charts. Table 4-1 illustrates and describes several of the charts available to you.

Table 4-1 Chart Types

Chart	Description
	XY (scatter) charts show relationships between two ranges of numeric data.
	Pie charts identify the relationship of each value to the entire data range.
	Column charts compare individual values or sets of values.
	Stacked column charts compare individual and total values by stacking columns on top of each other in a single column.
	Line charts plot changes in data. Each line represents a data range; each point on the line represents a value in the data range.

This chapter begins by creating an *XY* (scatter) chart that represents the data from a worksheet designed to store and filter digital signals. Next you learn how to create pie charts, bar charts, and line charts using the data generated by circuit boards analysis. Finally, a section summarizes methods for importing data files from other software packages or languages into Excel so you can generate charts for these data files.

4-1 CREATING *XY* CHARTS

An *XY (scatter) chart*, which plots data points against a numeric *x*-axis, is the most commonly used graph for engineering and scientific applications. This section describes how to create, view, enhance, name, and print *XY* charts.

Creating a Simple Chart

When you finish entering a set of data into a worksheet, or when you calculate a set of data using worksheet computations, you often want to generate a simple chart of the data quickly to see if it appears to be correct. You will use the FILTER1 worksheet shown in Figure 4-1 to create an *XY* chart.

Figure 4-1
FILTER1 Worksheet

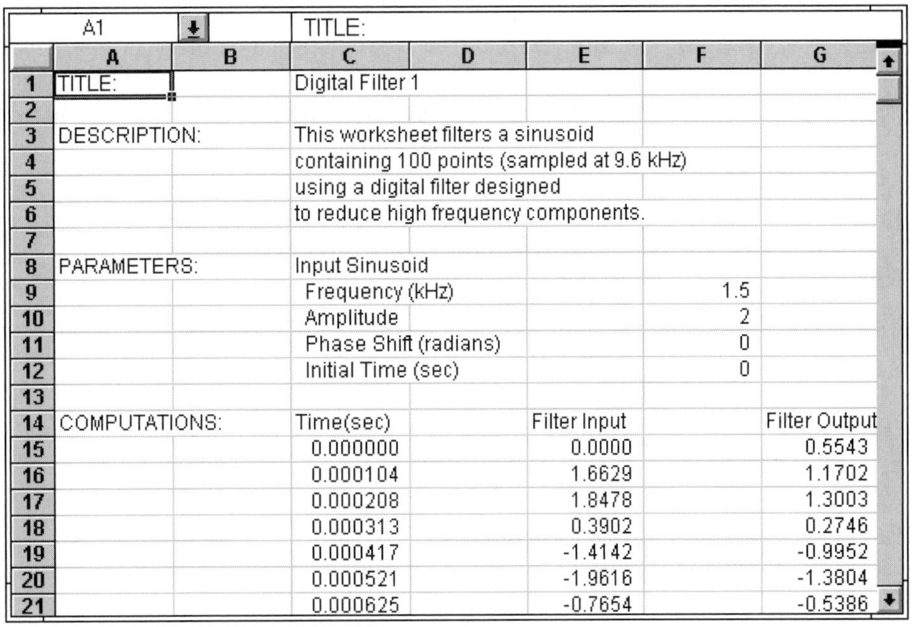

	A1	↕		TITLE:			
	A	**B**	**C**	**D**	**E**	**F**	**G**
1	TITLE:		Digital Filter 1				
2							
3	DESCRIPTION:		This worksheet filters a sinusoid				
4			containing 100 points (sampled at 9.6 kHz)				
5			using a digital filter designed				
6			to reduce high frequency components.				
7							
8	PARAMETERS:		Input Sinusoid				
9			Frequency (kHz)			1.5	
10			Amplitude			2	
11			Phase Shift (radians)			0	
12			Initial Time (sec)			0	
13							
14	COMPUTATIONS:		Time(sec)		Filter Input		Filter Output
15			0.000000		0.0000		0.5543
16			0.000104		1.6629		1.1702
17			0.000208		1.8478		1.3003
18			0.000313		0.3902		0.2746
19			0.000417		-1.4142		-0.9952
20			0.000521		-1.9616		-1.3804
21			0.000625		-0.7654		-0.5386

The worksheet was created by applying a digital filter to a sinusoid function, which is one of the most commonly used functions in digital signal processing. A sinusoid is a sine function written as a function of time; its general form is

$$x(t) = A \sin (\omega t + f)$$

where *A* is amplitude, *ω* is frequency in radians per second, *t* is time in seconds, and *f* is phase shift in radians.

To create an *XY* (scatter) chart, you use the ChartWizard button which is shown in Figure 4-2, and is found on the standard toolbar. You can either put the chart on the same sheet as your data or on a new sheet. When you click the ChartWizard button, the mouse pointer changes to a chart symbol. If you click anywhere in the workbook, Excel places the chart at that location. If you want to define the boundaries of your chart, you can drag the mouse.

Figure 4-2
ChartWizard Toolbar Button

You create a new chart by using a series of five ChartWizard dialog boxes. With the first dialog box, you select the data ranges to use. You use the second dialog box to select the chart type. With the third dialog box, you select the chart format. The fourth dialog box allows you to see a preliminary chart and add embellishments. Finally, you use the fifth dialog box to add labels to your chart. Each dialog box gives you the option to return to the previous step or to cancel your chart.

To illustrate, the first step is to select the data ranges C15:C114 and E15:E114 from the worksheet for the chart. Click the ChartWizard button, and the ChartWizard Step 1 dialog box appears, displaying the range addresses as shown in Figure 4-3. You could also have typed range names in the dialog box. Click the Next button to continue.

Figure 4-3
**ChartWizard Step 1
Dialog Box**

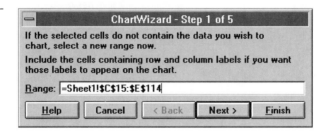

To plot the input to the digital filter (data in the Filter Input column of the worksheet), you need to change the chart type to an *XY* chart by choosing the *XY* (scatter) icon from the fifteen options given in the Step 2 dialog box shown in Figure 4-4. If you click the *XY* (scatter) icon and then click the Next button, the Step 3 dialog box displays chart icons representing the six formats available for *XY* charts, as shown in Figure 4-5. The default format puts data points on a Cartesian grid. For this example, click the lines and points format icon, and then click the Next button.

Figure 4-4
**ChartWizard Step 2
Dialog Box**

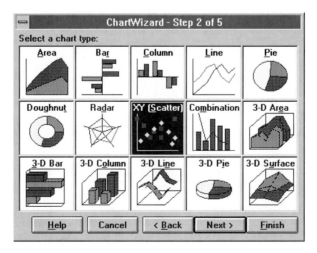

The Step 4 dialog box contains a sample chart of the specified data, as shown in Figure 4-6. Choose the Data Series in Columns button. Then choose the first column as the *x* data and the zero row as the Legend Text.

If you want to see a different format for your *XY* (scatter) chart, click the Back button. To continue, click the Next button.

Figure 4-5
ChartWizard Step 3 Dialog Box

Lines and points

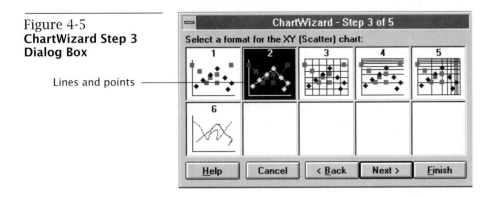

Figure 4-6
ChartWizard Step 4 Dialog Box

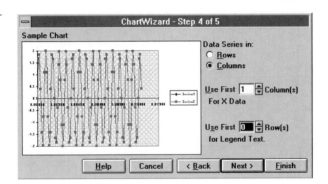

Finally, in the Step 5 ChartWizard dialog box shown in Figure 4-7, choose the No button under Add a Legend?, and fill in the Chart Title and Axis Title. Be sure to use the mouse to move from entry to entry, and do not hit (ENTER) until you have entered all the text. Then choose the Finish button. The chart now appears in your workbook along with a ChartWizard toolbar.

Figure 4-7
ChartWizard Step 5 Dialog Box

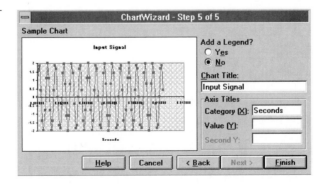

You might find that you would like to resize or move your chart. To resize your chart, select it and drag the mouse to resize it. To move your chart, select it, and use the Edit Cut command. Then select a cell elsewhere in the workbook, and use the Edit Paste command to display your chart there.

You also can use the ChartWizard toolbar, which has some basic built-in formatting capabilities to change the chart type, to change your chart to the default chart type, to change the range of values in the chart, to change the rows and columns used for labels, and to add and delete horizontal gridlines or a legend (the labels used to identify the different functions or sets of information compared on the graph).

Changing the Range

To change the ranges used in the chart, select the ChartWizard button. A ChartWizard Step 1 dialog box appears. Change the ranges to C14:C114 and E14:E114. Then click the Next button. Using the Step 2 dialog box, you can include the first row as the legend text. Click the OK button to exit the ChartWizard.

Enhancing a Chart

Microsoft Excel provides various *enhancements* you can use to improve the appearance of your charts. For example, you can change or delete chart titles and legends and you can add gridlines. Online Help describes additional options you can use as you prepare charts.

To demonstrate how to enhance a chart, you will change the *x*-axis title and add gridlines to the chart shown in Figure 4-8. To edit a chart, you first select it by double-clicking it. Deleting the chart's title is as simple as pointing to the title, clicking the left mouse button, and pressing (DEL). To change the chart's title, you first select it with the mouse. Then you type the new title. The title appears in the edit line as you type. Then press (ENTER). You can also delete the title and insert a new title using the Title command. Click the button for the type of titles you want to insert, as shown in Figure

Figure 4-8
Simple *XY* Chart of Filter Input

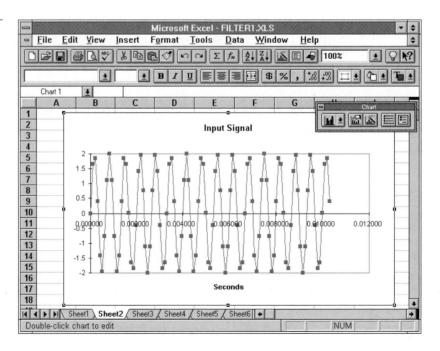

4-9. Then type the new title and press (ENTER). You can use the same technique to change the *x*-axis label.

Figure 4-9
Titles Dialog Box

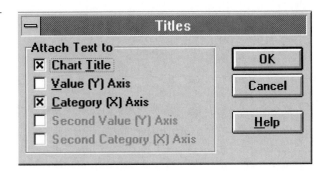

To add gridlines, choose the Insert Gridlines command. (If the chart is not selected, you will not see this command on the Insert menu.) The Gridlines dialog box appears and you can select major and minor gridlines in the *x* and *y* directions by checking the boxes. For horizontal gridlines only, you can use a ChartWizard toolbar button. The final enhanced chart is shown in Figure 4-10.

Figure 4-10
**Enhanced Chart of
Filter Input**

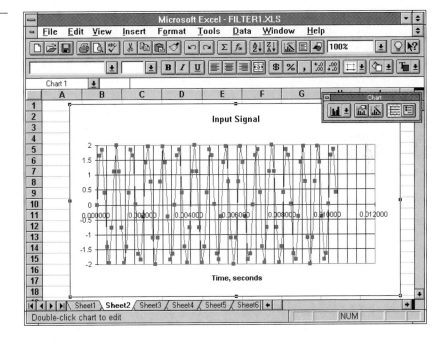

Try It

Make the indicated enhancements to the chart of the output signal of the FILTER1 worksheet, which is included in the data files for this module. View the plot after each enhancement to be sure that the change is what you expected.

◆ Change the title to Digital Filter.

◆ Change the *x*-axis label to Time, seconds.

◆ Change the *y*-axis label to Filter Output Sinusoid.

◆ Change the chart type to display only symbols for the data points. (Use the Format AutoFormat command.)

◆ Remove the gridlines.

◆ Scale the chart so that it fills the window.

Naming, Saving, and Printing a Chart

A worksheet can contain more than one chart. To simplify the process of locating charts in a large file, you can put them on different sheets. To help you remember the name and location of the chart you just created, use the Format Sheet Rename command to give the sheet a more meaningful name. Type the name in the text box. Then you can use the sheet tabs to locate the chart from any point in the worksheet.

Saving and printing charts is identical to saving and printing worksheets. The chart is automatically saved when you save the worksheet file. You can either print the chart by itself or with the worksheet. To print the chart by itself, select the chart sheet, and then choose File Print. In the Print dialog box, click the Selected Sheet button. Then click OK to start printing.

Try It

Change the name of the sheet in the FILTER1 worksheet to SIGNAL_IN. Then print the chart.

| **Application 1** | **QUALITY CONTROL** |

Manufacturing Engineering

In a manufacturing or assembly plant, quality control receives close attention. One of the key responsibilities of a quality control engineer is to collect accurate data on the quality of the product being manufactured. This data can be used to identify the problem areas in the assembly line or in the materials being used in the product.

Circuit Board Defects

In this application, information collected over a one-year period is used to specify both the type of defects and the number of defects detected in the assembly of printed circuit boards. These defects have been divided into four categories: board errors, chip errors, processing errors, and connection errors. Board errors are typically caused by defects in the wire traces of the printed circuits. Chip errors are caused by defective integrated circuit (IC) chips that are added to the board; these IC chips include memory chips, microprocessor chips, and digital filter chips. Processing errors are typically caused by errors in inserting the chips in the board; this process is often done by manufacturing robots, and the robot programming can be incorrect, or the chips being inserted can be packaged in the wrong order. Connection errors are solder errors that occur when the board goes through the solder machine; these errors can be caused by dirt on the board or an incorrect solder temperature.

Worksheet for a Quality Analysis Report

You want to develop a worksheet that summarizes the quality control data that has been collected each month for a year. This data includes the number of defects in each of the four categories of defects discussed. The summary report should compute totals and percentages for the entire year and defect totals for each quarter. Later in this chapter you will use the data in the worksheet to generate pie, bar, and line charts.

1. Problem Statement

Generate a summary of the quality control monthly data over the past year. Include totals and percentages of the four defect categories along with quarterly totals.

2. Input/Output Description

Begin the worksheet using a solution template containing a title, description, and the parameters of the worksheet. The monthly data represents the input parameters. The QUALITY worksheet shown in Figure 4-11 contains the title, description, and parameter area. You now need to specify the form of the quality analysis report that is generated by the worksheet. Figure 4-12 contains a window of the report template.

Figure 4-11
QUALITY Worksheet

	A1	↓		TITLE:					
	A	**B**	**C**	**D**	**E**	**F**	**G**	**H**	
1	TITLE:		Quality Control						
2									
3	DESCRIPTION:		This worksheet generates a quality analysis						
4			report using data collected in the manufacturing						
5			of printed circuit boards. The analysis covers						
6			one year, and includes total information						
7			and quarterly information on defects.						
8									
9	PARAMETERS:		Month	Boards	Boards		Numbers of Defects		
10				Passed	Rejected		Board	Chip	
11			Jan	1201	34		8	12	
12			Feb	890	15		3	5	
13			Mar	933	24		13	6	
14			Apr	1022	18		9	3	
15			May	975	10		4	3	
16			Jun	864	13		8	4	
17			Jul	891	17		6	5	
18			Aug	903	11		4	6	
19			Sep	1075	18		10	3	
20			Oct	1180	21		11	6	
21			Nov	1380	34		20	11	
22			Dec	903	11		4	2	
23									
24									

Figure 4-12
Report Template

	A	B	C	D	E	F	G	H
	A26	↓		ANNUAL QUALITY ANALYSIS REPORT				
26	ANNUAL QUALITY ANALYSIS REPORT							
27								
28	Total Number of Boards Produced =							
29	Non-Defective Boards =							
30	Defective Boards =							
31								
32	Defect Analysis:							
33								
34	Defect Type			Count	Percentage			
35	Board							
36	IC Chip							
37	Process							
38	Solder							
39								
40	Quarterly Defect Analysis:							
41								
42	Quarter	Board	Chip	Process	Solder	Total		
43		Defects	Defects	Defects	Defects	Defects		

3. Hand Example

Generate a sample set of data:

	Boards Passed	Boards Rejected	Number of Defects			
			Board	Chip	Process	Solder
Jan	1201	34	8	12	4	10
Feb	890	15	3	5	2	5
Mar	933	24	13	6	0	5
Apr	1022	18	9	3	3	3
May	975	10	4	3	1	2
Jun	864	13	8	4	0	1
Jul	891	17	6	5	2	4
Aug	903	11	4	6	0	1
Sep	1075	18	10	3	4	1
Oct	1180	21	11	6	1	3
Nov	1380	34	20	11	0	3
Dec	903	11	4	2	3	2

Using this information to compute by hand the values for the summary report, you get

Total Number of Boards Produced =		12,443
Non-defective Boards =	12,217	98.18%
Defective Boards =	226	1.82%

Defect Analysis:

Defect Type	Count	Percentage
Board	100	44.25%
IC Chip	66	29.20%
Process	20	8.85%
Solder	40	17.70%

Quarterly Defect Analysis:

Quarter	Board Defects	Chip Defects	Process Defects	Solder Defects	Total
I	24	23	6	20	73
II	21	10	4	6	41
III	20	14	6	6	46
IV	35	19	4	8	66

4. Algorithm Development

The worksheet now contains everything except the formulas for computing the summary information for the report. Develop the formulas in the order needed to compute the values. Also, try to minimize the number of computations. For example, since you generate error sums by quarters, add the quarterly sums to get yearly sums instead of adding all the monthly sums to get yearly sums. It would be good practice to verify each of these formulas by referring to Figures 4-11 and 4-12.

D29	=SUM(D11:D22)	Total non-defective boards
D30	=SUM(E11:E22)	Total defective boards
E28	=D29+D30	Total boards
E29	=D29/$E$28	Percent non-defective boards
E30	Copy from E29	Percent defective boards
B44	=SUM(G11:G13)	Quarter I board defects
B45	=SUM(G14:G16)	Quarter II board defects
B46	=SUM(G17:G19)	Quarter III board defects
B47	=SUM(G20:G22)	Quarter IV board defects
C44:E47	Copy from B44:B47	Other quarter defect sums
F44	=SUM(B44:E44)	Total Quarter I defects
F45:F47	Copy from F44	Other quarter total defect
D35	=SUM(B44:B47)	Total board defects
D36	=SUM(C44:C47)	Total IC chip defects
D37	=SUM(D44:D47)	Total process defects

D38	=SUM(E44:E47)	Total solder defects
E35	=D35/$D$30	Compute percent board defects
E36:E38	Copy from E35	Compute other percent defects

The appearance of some of the information computed for the report could be improved by formatting. Use the Format Cells Number command to specify the following formatting:

Format	Cells
Comma, 0 decimal places	E28, D29, D30, D35:D38
Percent, 2 decimal places	E29, E30, E35:E38

You can also name the following ranges:

TYPES	A35:A38
TOTALS	D35:D38
QTR1	B44:E44
QTR2	B45:E45
QTR3	B46:E46
QTR4	B47:E47
REPORT	A26:F47

These will be useful in developing charts in the next section and in printing the summary report.

5. Testing

An important part of developing a worksheet is testing it with several sets of data to verify the accuracy of the computations. Using the sample set of data from the hand example, you can easily check the accuracy of the worksheet calculations by comparing them to the hand example.

You should make minor changes in this data and check the report to be sure that corresponding changes occur in the report summary. For example, it is important to check the limits of the values. In this report you want to be sure that the report would be generated correctly if there were no errors in one of the categories. Therefore, you can change the total processing defects for each month to zero to be sure that the summary report would still be correct. Be sure to change the corresponding sums of boards rejected. The corresponding report generated, shown in Figure 4-13, shows that there were no process defects during any of the four quarters. Also, note that the total number of defects was reduced by 20 (there were 20 processing defects in the original data set).

In this section you have developed a worksheet to compute numerical quality control information; in the next section you will use pie charts, bar charts, and line charts to display this information. As a prelude to this next section, view some of the charts included in the QUALITY workbook by clicking the sheet tabs. All of these charts will be developed in the next section.

Figure 4-13
**Report with No
Processing Defects**

A26	↓		ANNUAL QUALITY ANALYSIS REPORT					
	A	B	C	D	E	F	G	H
26	ANNUAL QUALITY ANALYSIS REPORT							
27								
28	Total Number of Boards Produced =							
29	Non-Defective Boards =							
30	Defective Boards =							
31								
32	Defect Analysis:							
33								
34	Defect Type			Count	Percentage			
35	Board							
36	IC Chip							
37	Process							
38	Solder							
39								
40	Quarterly Defect Analysis:							
41								
42	Quarter	Board	Chip	Process	Solder	Total		
43		Defects	Defects	Defects	Defects	Defects		

What If

Assume that the assembly line was closed for the month of August. Set the August totals to zero, and check the resulting report.

Assume that the assembly line was closed for the entire third quarter of the year. Set the appropriate monthly totals to zero, and check the resulting report.

4-2 CREATING OTHER TYPES OF CHARTS

Although the *XY* chart is the most commonly used chart type for engineering and scientific problem solutions, other chart types are also useful. This section uses the data generated in the QUALITY workbook to demonstrate how to develop pie charts, bar charts, and line charts. The discussion of each type of chart focuses on two issues: the ease of generating an initial chart and the options available for enhancing the chart to a final form. You can select the chart's ranges from the worksheet before clicking the ChartWizard button. Microsoft Excel then plots the data series for you according to the following rules:

- If the selected range contains more rows than columns, Excel plots the data series by columns.
- If the selected range contains more columns than rows, Excel plots the data series by rows.
- Excel does not count blank columns or rows contained in the range.

Pie Charts

A *pie chart* is useful for displaying the different categories of an item. For example, the previous section discussed the possible defects that can occur in the assembly of a printed circuit board. If you are analyzing the defects discovered in the completed circuit boards, you may be interested in a pie chart that displays not only the different defects that occur but also the percentage of occurrence of the various defects.

To demonstrate the ease of generating a pie chart, use the QUALITY worksheet from the last section as your current worksheet. Begin by selecting the range TOTALS. Then choose the ChartWizard button and put the chart on its own sheet. Since the range is already correct, you don't need to type anything for the Step 1 dialog box; just click the Next button. Use the Step 2 dialog box to change the default chart to a pie chart. Click the pie chart icon, and then click the Next button. Since you want to use the default pie chart format, click the Next button on the Step 3 dialog box. The pie is displayed in the Step 4 dialog box. Click the Next button again. Then click the Finish button on the Step 5 dialog box to finish the pie chart, which is shown in Figure 4-14. Rename the sheet PIE1.

Figure 4-14
Simple Pie Chart

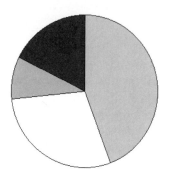

You can enhance the pie chart by adding labels to the slices and adding a title. To label the pie slices using the named range TYPES, you choose Insert New Data, and enter TYPES as the Range. To add a title, choose Insert Title. Then double-click the default title, and type Annual Summary of Defect Types in the edit box. Name the enhanced pie chart PIE2; it is shown in Figure 4-15. If you want to change the colors or add patterns, double-click the pie slice and click the desired color from the palette.

Figure 4-15
Enhanced Pie Chart

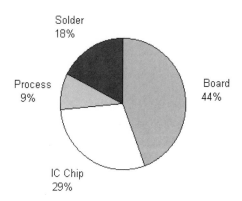

Annual Summary of Defect Types

One of the options available with pie charts allows you to select a piece of the pie and separate it from the rest of the pie. The slice that is separated is called an *exploding slice.* You can explode one or all of the slices by selecting the slice and dragging it away from the other slices. If you are particularly interested in the IC chip errors, you could use this technique to emphasize chip errors while still displaying the information on the other defect types. Figure 4-16 shows the enhanced pie chart with the IC chip category separated.

Figure 4-16
**Enhanced Pie Chart
with Exploding Slice**

Annual Summary of Defect Types

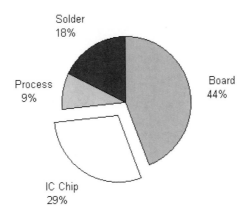

Column Charts

Another type of chart that can be very useful for illustrating worksheet values is the default bar chart, called a *column chart* in Excel. In this section you will create a column chart containing the defect types. The *y*-axis will be TOTALS, and the *x*-axis label will be TYPES. Begin by selecting these ranges and clicking the ChartWizard button. Select a new sheet for the chart. The simple column chart, named BAR1, is shown in Figure 4-17.

Figure 4-17
Simple Column Chart

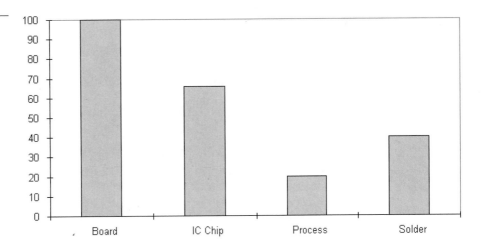

You can enhance the chart by changing the type to a 3-D column chart. To change the simple column chart to a 3-D column chart, either select the 3-D column chart icon shown in Figure 4-18 from the ChartWizard toolbar, or use the Format AutoFormat command. If you want to add or change any of the titles, edit them by selecting them and then typing in the new title. The enhanced chart named BAR2 is shown in Figure 4-19. Note that the column chart usees the actual data values from the worksheet (number of defects) instead of computing percentages from the values as done for pie charts.

Figure 4-18
3-D Column Chart Icon

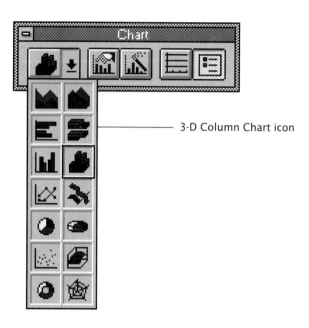

3-D Column Chart icon

Figure 4-19
Enhanced Column Chart

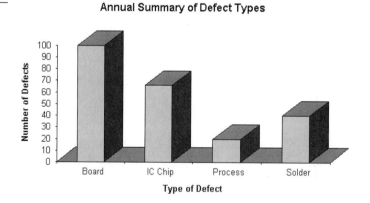

If you want to compare the defect types over the four quarters, Excel offers two different options with column charts. The chart in Figure 4-20 is a column chart, and the chart in Figure 4-21 is a *stacked column chart*. The column chart is named BAR3, and the stacked column chart is named BAR4. Both charts contain the same information, but they display it differently.

Figure 4-20
Column Chart

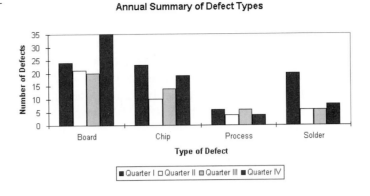

Figure 4-21
Stacked Column Chart

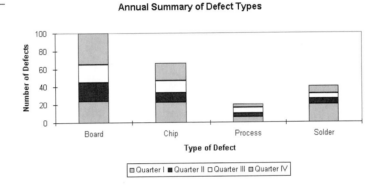

To create the column chart in Figure 4-20, first select the range A42:E47 with the mouse, and then click the ChartWizard button. The column and row headings will automatically be used in your chart if you choose the first two rows as the *x*-axis label and the first column as the legend text. Also, be sure to choose the data series in rows. To move the legend, double-click on the legend and select the Placement tab in the Format Legend dialog box. To change the legend entries, select the corresponding data on the chart. For example, click on a column in the chart that represents board defects. Notice that all of the columns for that type of defect are highlighted. Double-click the column, and select the Names and Values tab from the Format Data Series dialog box. Then change the Name field to Quarter I. Repeat for the other quarters. The resulting chart is shown in Figure 4-20.

Changing the column chart to a stacked column chart is easy. Choose the Format AutoFormat command. Ten types of column charts are displayed. To change the column chart to a stacked column chart, select the icon of the stacked column chart and click OK. The resulting chart is shown in Figure 4-21.

Line Charts

At first glance, line charts seem to be the same as *XY* charts, but there is a distinct difference. *XY* charts show numerical relationships using numerical values for both the *x*-axis and the *y*-axis. *Line charts* show the numerical relationship for only y-axis values; the x-axis values are literals. Thus the x-axis literals can be board defect type (Board, IC Chip, Process, Solder), or they might be years that correspond to population values (1960, 1970, 1980, 1990). It is usually not difficult to decide whether you want an XY chart or a line chart. If you have xy coordinates, then you probably want an XY chart; if you have only a set of numbers and their corresponding labels, then you probably want a line chart.

You can create a simple line chart containing the defect types using the QUALITY worksheet. First click the ChartWizard button, and enter TOTALS and TYPES as the ranges in the ChartWizard Step 1 dialog box. Click the Next button. Use the Step 2 dialog box to change the default chart to a line chart by clicking the line icon. Then click the Next button. Use the Step 3 dialog box to select a line format as well, and click the Next button. If you don't select a legend in the Step 4 dialog box, click the Next button and then the Finish button in the Step 5 dialog box. The simple line chart in Figure 4-22 appears. This chart is named LINE1.

Figure 4-22
Simple Line Chart

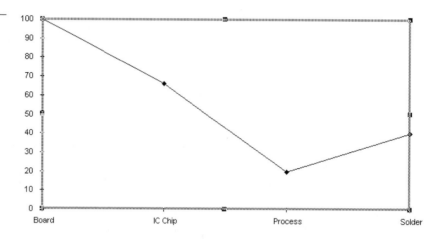

You can enhance this chart as you did for the pie charts and bar charts. You can also use the line chart to compare the defect types by quarters, using essentially the same steps you used when comparing defect types by quarters with a column chart (see Figure 4-20). Select the following ranges: QTR1, QTR2, QTR3, and QTR4. Click the ChartWizard button, and then click the Next button in the ChartWizard Step 1 dialog box. In the Step 2 dialog box, select the line chart icon again, and in Step 3 select the first format style. Click the next button in Step 4. Click the legend option button in the Step 5 dialog box. Add the chart title Annual Summary of Defect Types, an x-axis label of Type of Defect, and a y-axis label of Number of Defects. After you have clicked the Finish button, a chart is displayed, but the legend needs editing. To change the legend, double-click the data on the chart corresponding to Series I in the legend. Then using the Names and Values tab in the Format Data Series dialog box, change the name that will appear in the legend from Series I to Quarter I. Repeat this process for each line in the chart. Then, if you add horizontal gridlines with the ChartWizard toolbar (or choose Insert Gridlines Major Intervals for the *Y*-axis), the line chart shown in Figure 4-23 is displayed. This chart is named LINE2.

Figure 4-23
**Line Chart
Comparisons**

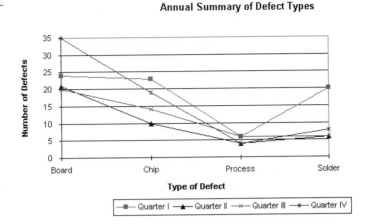

You can switch from one type of chart to another by using the ChartWizard toolbar or the Format AutoFormat menu. For example, you can change the line chart in Figure 4-23 to a column chart by choosing Format AutoFormat, selecting Column, and removing the horizontal grid lines. Note that you can often get the same results even faster using the ChartWizard toolbar.

You have learned how to create the charts most commonly used in engineering and scientific problem solutions—*XY* (scatter) charts, pie charts, bar charts, and line charts. With Microsoft Excel you can also create HLCO (high-low-close-open) charts to track changes in data over time and mixed charts that combine lines and bars in the same chart. Take time to experiment with these charts and with other chart options available in the dialog boxes. You will find that the charting capabilities of Excel are very powerful because you can switch easily between chart types and options to find the settings that best display your data.

Try It

◆ Explode a different slice in PIE2.

◆ Use 3-D bars in BAR4.

◆ Remove the horizontal grid lines from LINE2.

◆ Identify the toolbar buttons to create pie and line charts. Use these toolbar buttons to change BAR2 to a pie and then to a line chart.

4-3 IMPORTING DATA FILES

Since different software packages and programs have different capabilities, you might want to use more than one package or program to manipulate the same data. For example, you may have a Fortran program that generates a data file and you want to plot the information in the file. A good solution would be to import the data file into an Excel worksheet and then use the Chart command and its various options to generate the chart. You may also have data generated by the other spreadsheets or word processors that you want to incorporate into a worksheet for further analysis or plotting. The next section summarizes various ways of importing data into Excel worksheets.

Importing Files

The File Open command copies data from a text file, also called an ASCII (American Standard Code for Information Interchange) file. The text file should be saved with the extension .TXT.

There are two types of text files, delimited and nondelimited. A *delimited file* separates labels and numbers by commas, spaces, colons, or semicolons. In addition, labels are enclosed in quotation marks. Most computer languages generate delimited files unless the program specifies another format. The data in a *nondelimited file* is not separated. However, each line in the file must end with a carriage return or a line feed and must not exceed 240 characters. Most word processors store text in a nondelimited form if you specify a text file when you save the file. (Word processors do not generally store files as text files; you must specifically select a text file format for information that you plan to use with other software.)

You import both delimited and nondelimited files by choosing the File Open command. The Open dialog box shown in Figure 4-24 appears. Select Text (Files) in the file type drop-down list. Then select the file you want in the File Name text box.

Figure 4-24
Open File Dialog Box

File type
drop-down list

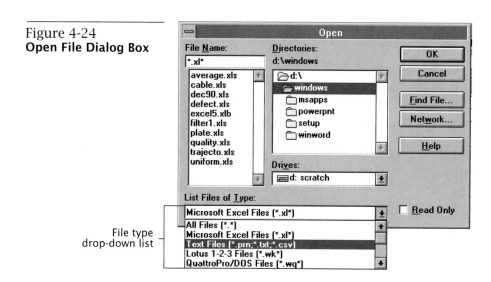

To store a text file in a worksheet, it is necessary to use the Text Import Wizard that leads you through a series of three dialog boxes. The options vary depending on whether your data is delimited or not. After each step click the Next button. After the final step click the Finish button. The text file is then automatically stored in the worksheet. Each text entry is placed in a separate cell beginning with the cell A1; with each new line in the data file, the current cell moves to the next row. If you import a delimited file and select no delimiters, each line in the text file is stored in a single cell in the worksheet. The Text Import dialog boxes for delimited data are shown in Figures 4-25 through 4-27.

Figure 4-25
**Text Import Wizard
Step 1 Dialog Box**

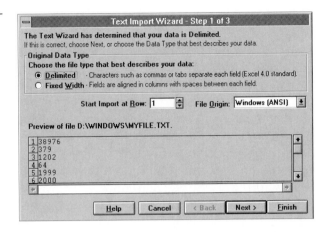

Figure 4-26
**Text Import Wizard
Step 2 Dialog Box**

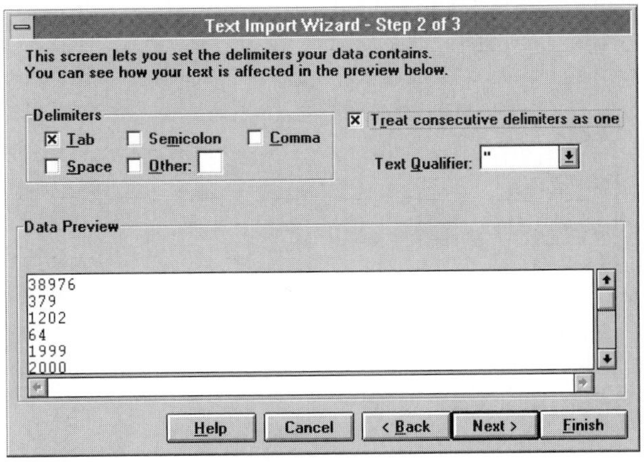

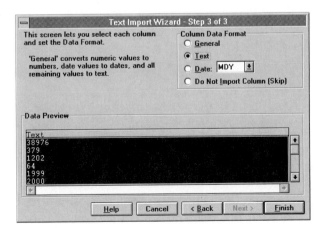

Figure 4-27
**Text Import Wizard
Step 3 Dialog Box**

Parsing Imported Data

In addition to the Text Import Wizard options for formatting imported data, the Data Text to Columns command converts the long labels imported by the File Open command into separate columns of data. Since this command is seldom needed to incorporate data into a worksheet, it is not covered in detail here, but information is included in online Help. Search for Parsing Data and then go to Converting Text to Columns.

SUMMARY

This chapter demonstrated how to create *XY* (scatter) charts, pie charts, bar charts, and line charts. The first step is to quickly generate a simple chart to view the data and verify its accuracy. Changing the chart type allows you to view this data in several different forms and to decide the best way to represent it. Then, once you have determined the chart type, you have a number of options for enhancing the chart and adding critical information. If you assign meaningful names to charts, you can locate them quickly using the Edit Go To command. This chapter also showed how to import files generated outside Excel into worksheets for further analysis and for generating charts.

Key Words

column chart	nondelimited file
delimited file	pie chart
enhancement	stacked column chart
exploding slice	*XY* (scatter) chart
line chart	

Exercises

These exercises involve modifications to the worksheet generated in this chapter and to worksheets from previous chapters. Start each exercise with the original worksheet. Add enhancements to any plots that you create.

QUALITY Worksheet

1. Generate a pie chart showing the defect types for the month of November.

2. Generate a bar chart comparing the monthly defect types for the months of January through April.

3. Generate a line chart showing the number of boards rejected by month for the year.

4. Generate a line chart with two lines: one showing the number of boards rejected by month for the year and one showing the number of boards passing inspections by month for the year.

5. Modify the worksheet to contain the number of defects by month, and then generate a line chart showing this information.

UNIFORM Worksheet

6. Add a named chart to the worksheet that generates a line plot of the random numbers. (Why is this not an *XY* chart?)

CABLE Worksheet

7. Add a named chart to the worksheet that generates an *XY* chart of the velocity of the cable car.

FILTER1 Worksheet

8. Generate an *XY* chart showing the difference between the input to the filter and the output from the filter.

TRAJECTORY Worksheet

9. Generate an *XY* chart showing the altitude data in feet instead of meters.

AVERAGE Worksheet

10. Plot the experimental data using a line chart.

11. Add a horizontal line to the line chart in problem 10 that shows the average value of the experimental data.

New Spreadsheets

12. Develop a spreadsheet that contains x and y values of the following cubic (third-degree) polynomial:

$$y = f(x)$$
$$= x^3 - 5x^2 + 2x + 8$$

The x values should begin at –2 and increment by 0.1 through the value 5.0. Generate an XY plot of these values.

13. Modify the spreadsheet developed in problem 12 so that it has parameters for the beginning x value and for the increment between the x values. Assume that there will always be 100 values of x used.

14. A root of an equation y = f(x) is a value of x that corresponds to a function value of zero. From the plot generated in problem 12, it is clear that the roots of this polynomial are –1, 2, and 4. Modify the spreadsheet in problem 12 so that it identifies lines of the report that correspond to a root with the label root. Thus, the line in the report that corresponds to the value –1 should have this format:

–1.0   0.0   root

The spreadsheet should determine the roots by comparing the absolute value of the function value to 0.00001. If the absolute value of the function value is less than this value, then the label root should be added to the line.

15. Modify the spreadsheet in problem 14 so that it generates values of a general cubic equation. Thus, parameters in the spreadsheet determine the value a, b, c and d of the general cubic equation:

$$y = f(x)$$
$$= ax^3 + bx^2 + cx + d$$

5 Scientific Databases

Meteorology  Meteorology is the science that deals with the atmosphere and its phenomena, and especially with weather and weather forecasting. Technology has aided meteorologists in the prediction of weather with satellites that collect images worldwide and with remote sensing stations that constantly relay weather information. In addition, technology has developed new tools such as Doppler radar, which converts the shift in frequency of moving targets into speed for measuring wind speed and direction. A great deal of information about weather can also be learned from analyzing past and present data to observe the conditions that lead to extreme weather events such as tornadoes, hurricanes, typhoons, and microbursts. A database is an ideal structure to store, analyze, and display weather information. In this chapter you will develop a database using Excel lists to analyze historical weather data.

INTRODUCTION

In Chapters 1 through 3 you learned how to use the spreadsheet features of Excel to store data in worksheets and then solve engineering and science problems concerning the data. But Excel is an integrated program that provides additional capabilities, and in Chapter 4 you learned how to use the charting feature of Excel to display worksheet data in various charts. In this chapter you will learn how to use Excel lists to analyze worksheet data.

When using Excel as a spreadsheet program, you create formulas to perform computations on the worksheet data. When using Excel as a database program, you use the database commands and functions to analyze worksheet data that is organized in columns and rows. For example, you can select data that satisfy certain *criteria*, such as locating the fastest time or the smallest distance, or you can rearrange the data in different ways, such as in ascending or descending order. You also can perform statistical calculations on the data, such as finding the average or variance of a set of values or even creating a frequency distribution. Finally, you can design and print reports that display the results.

A *database* is a collection of information that can be organized into list form. In Excel, a database is an area of a worksheet that is organized into rows and columns. Excel refers to a database as a *list* and database operations as list management operations. The commands presented in this chapter are Data commands. While we are discussing them in the context of their usefulness with a list, some of these commands can be used with data that is not formally a list.

This chapter discusses creating a list and illustrates the steps using circuit board defect data and climatology data from the National Weather Service. The commands for sorting a list are presented. A discussion of criteria for selecting information from a list follows, along with updating or deleting the information and creating filtered lists. This chapter also discusses a number of Excel functions that perform computations on the entire list or on subsets of the list.

5-1 CREATING A DATABASE USING AN EXCEL LIST

In Excel a list is organized into rows and columns. Each row in a list is a *record*, and each column in a list is a *field*. Thus each record contains a group of fields. To help you visualize a list, this section uses the defect information from the quality control application in Chapter 4. Since a list containing this type of information might include data for several years, the data from Chapter 4 is expanded by adding two more years of data and an extra field that contains the year.

Specifying Fields and Records

The Excel list management commands and functions use a *field name* to refer to fields within the records. This field name must be unique for each field and is stored at the top of each column. The field name must begin with a letter or an underscore. The other characters can be letters, numbers, periods and underscores. The field name can be up to 255 characters long.

Table 5-1   Three Years of Defect Information

Date			Total		Defect Type		
Year	Month	Passed	Rejected	Board	Chip	Process	Solder
89	Jan	1201	34	8	12	4	10
89	Feb	890	15	3	5	2	5
89	Mar	933	24	13	6	0	5
89	Apr	1022	18	9	3	3	3
89	May	975	10	4	3	1	2
89	Jun	864	13	8	4	0	1
89	Jul	891	17	6	5	2	4
89	Aug	903	11	4	6	0	1
89	Sep	1075	18	10	3	4	1
89	Oct	1180	21	11	6	1	3
89	Nov	1380	34	20	11	0	3
89	Dec	903	11	4	2	3	2
90	Jan	1151	25	6	9	2	8
90	Feb	972	13	3	4	2	4
90	Mar	863	18	10	6	0	2
90	Apr	1010	14	8	2	3	1
90	May	1002	9	3	3	1	2
90	Jun	907	12	8	3	0	1
90	Jul	983	18	7	5	2	4
90	Aug	956	16	6	9	0	1
90	Sep	1071	21	13	3	4	1
90	Oct	835	15	8	4	1	2
90	Nov	762	12	10	2	0	0
90	Dec	741	14	6	3	5	0
91	Jan	910	18	4	2	4	8
91	Feb	989	17	5	5	2	5
91	Mar	1010	23	12	6	0	5
91	Apr	983	19	10	3	3	3
91	May	868	9	4	3	0	2
91	Jun	750	12	8	4	0	0
91	Jul	802	12	3	3	2	4
91	Aug	792	9	4	4	0	1
91	Sep	918	14	10	1	2	1
91	Oct	952	18	8	6	1	3
91	Nov	1015	14	8	3	0	3
91	Dec	831	13	5	3	3	2

Table 5-1 contains defect information covering a period of three years. To analyze this information using database commands, it must first be stored in rows and columns in a worksheet. (This data is already stored in the DEFECT worksheet, which is included in the data files for this module.) The field names selected for this list are YEAR, MONTH, PASSED, REJECTED, BOARD, CHIP, PROCESS, and SOLDER. Figure 5-1 contains a window of the DEFECT worksheet containing this list. Note that, in addition to a title and description, a definition of each field is included in the list at the beginning of the worksheet. The widths of the data columns

Figure 5-1
DEFECT Spreadsheet

	A	B	C	D	E	F	G	H	I
1	TITLE:			Printed Circuit Board Defect Information					
2									
3	DESCRIPTION:			This database contains information on printed					
4				circuit board defects collected monthly					
5				during 1989, 1990, and 1991.					
6									
7	FIELDS:			RECORD - record number					
8				YEAR,MONTH - year and month of data collection					
9				PASSED - number of boards passing inspection					
10				REJECTED - number of boards rejected					
11				BOARD - number of boards rejected for board defects					
12				CHIP - number of boards rejected for IC chip defects					
13				PROCESS - number of boards rejected for processing defects					
14				SOLDER - number of boards rejected for solder defects					
15									
16	RECORD	YEAR	MONTH	PASSED	REJECTED	BOARD	CHIP	PROCESS	SOLDER
17	1	89	Jan	1201	34	8	12	4	10
18	2	89	Feb	890	15	3	5	2	5
19	3	89	Mar	933	24	13	6	0	5
20	4	89	Apr	1022	18	9	3	3	3
21	5	89	May	975	10	4	3	1	2
22	6	89	Jun	864	13	8	4	0	1
23	7	89	Jul	891	17	6	5	2	4

have been modified using the Format Column Width command in order to make it easier to read the information. When you work with a list, you frequently need to refer to the set of records within the list and to the set of records plus the field names. It is convenient to assign names to these two ranges:

DATA   A17:I52
INPUT   A16:I52

Note that INPUT refers to both the records (or DATA range) plus the field names.

Viewing Titles

Worksheets containing lists are often quite large. Since the field name labels are at the top of the list information, it can be difficult to remember the order of the information in the columns as you scroll down the list. The Window Freeze Panes command allows you to freeze particular rows or columns as titles, so that this information stays on the screen as you scroll through the data. To freeze titles for columns, position the titles that you want to freeze at the top of the window. Select the cells one row below the titles you want to freeze, and choose the Window Freeze Panes command. Figure 5-2 contains a window near the end of the DEFECT list with the field name labels displayed. Choose the Window Unfreeze Panes command to return to the regular display of information.

Displaying Record Numbers

It is often useful to display a record number next to each record, as shown in the DEFECT worksheet in Figures 5-1 and 5-2. You start by entering the first two numbers of a series in the first two rows of a column. Then select the two cells and drag the Fill Handle (black square located at the lower-

Figure 5-2
Window with Frozen
Panes of Row Titles

	A	B	C	D	E	F	G	H	I
	A16			RECORD					
16	RECORD	YEAR	MONTH	PASSED	REJECTED	BOARD	CHIP	PROCESS	SOLDER
31	15	90	Mar	863	18	10	6	0	2
32	16	90	Apr	1010	14	8	2	3	1
33	17	90	May	1002	9	3	3	1	2
34	18	90	Jun	907	12	8	3	0	1
35	19	90	Jul	983	18	7	5	2	4
36	20	90	Aug	956	16	6	9	0	1
37	21	90	Sep	1071	21	13	3	4	1
38	22	90	Oct	835	15	8	4	1	2
39	23	90	Nov	762	12	10	2	0	0
40	24	90	Dec	741	14	6	3	5	0
41	25	91	Jan	910	18	4	2	4	8
42	26	91	Feb	989	17	5	5	2	5
43	27	91	Mar	1010	23	12	6	0	5
44	28	91	Apr	983	19	10	3	3	3
45	29	91	May	868	9	4	3	0	2
46	30	91	Jun	750	12	8	4	0	0
47	31	91	Jul	802	12	3	3	2	4
48	32	91	Aug	792	9	4	4	0	1
49	33	91	Sep	918	14	10	1	2	1
50	34	91	Oct	952	18	8	6	1	3
51	35	91	Nov	1015	14	8	3	0	3
52	36	91	Dec	831	13	5	3	3	2

right corner of the selection) down the column. When you release the Fill Handle, the selected cells are filled with the next numbers in the series. The Fill Handle is not restricted to lists, so it can be used anytime that you need to fill a column or row with values that can be defined in terms of a starting number and an increment.

Try It Select an unused area of a worksheet and use the Fill Handle to generate the following information:

- ◆ A column of values that contains the numbers from 1 to 100.
- ◆ A column of values that contain the numbers from –50 to 50 in steps of 5.
- ◆ A row of 20 values that begins with 0.00 and increases in steps of 0.01.

Application 1   CLIMATOLOGY DATA

Environmental Engineering

A number of national agencies are interested in weather information. NOAA (National Oceanic and Atmospheric Administration) is a research-oriented organization that studies the oceans and the atmosphere. It also funds environmental research in data analysis, modeling, and experimental work relative to global changes. The National Environmental Satellite, Data, and Information Service collects and distributes information relative to the weather. The National Climatic Data Center collects and compiles climatology information from National Weather Service offices across the country. It is also the National Weather Service offices that interact with state and local weather forecasters to keep the general public aware of current weather information. This database application focuses on the type of information collected by the National Weather Service.

The National Climatic Data Center in North Carolina is responsible for maintaining climatological data from National Weather Service offices. This data is available in many forms, including local climatology data by month, data by state, and data for the world. It also maintains historical climatology data beginning with 1931. Figure 5-3 contains a monthly summary of local climatology data that was collected by the National Weather Service office at Stapleton International Airport in Denver, Colorado, for the month of December 1990. The summary contains 23 different pieces of weather information collected for each day, including maximum and minimum temperature, amount of precipitation, peak wind gusts, and minutes of sunshine. This data is then analyzed to generate the monthly summary information at the bottom of the form, which includes average temperature, total rainfall, total snowfall, and number of days that were partly cloudy.

This data clearly fits the requirements for a database and can be represented as a list—it is composed of records of information that contain individual fields of information. A record corresponds to the complete information collected for one day, and a field corresponds to one of the pieces of information collected during that day. As you will see later in this chapter, the results presented in the monthly summary are easily computed using database commands and functions. In this section you will generate a climatology list from the data for December 1990 and then use it to illustrate database commands and functions.

Creating a Worksheet for a Climatology Database

Creating a database requires careful consideration of several issues. You must consider how much descriptive information to include, the number of fields, the types of data in the fields, and the field names for the fields. The following section outlines these issues.

Descriptive Information

You can assume that this climatology database typically will be copied into a worksheet that will access it, or it will be combined with other databases. Since it will have a variety of uses, it is important to have a clear description of the fields and the types of information (or codes) stored in the fields.

Although you are generating a database of only one month's data, you can assume there will be many such databases for other months, and the databases will frequently be combined to generate databases covering larger periods of time. It would be redundant to include the discussion of the fields in each database; it is more efficient to generate one worksheet that contains all this information and to refer to this worksheet in each of the individual databases. Store the following information in a database in a worksheet entitled DEC90.

A1   TITLE:
C1   Climatology Data for December 1990
A3   DESCRIPTION:
C3   This database contains the climatology data

C4   collected at Stapleton International Airport
C5   for the month of December 1990.
A7   FIELDS:
C7   See the CLIMATE worksheet for a
C8   description of the fields and their codes.

When a database is one of a kind (unlike our example), a description of the fields should probably be stored at the beginning of the database unless it is very long; if the information is long, then a reference to the location of the information must be included in the description.

Number of Fields

Look carefully at Figure 5-3 to discover how many fields the list requires. The 23 fields that are numbered across the top of the summary in the figure will not be sufficient. For one thing, the DEGREE DAYS information is labeled as 7A and 7B, and will require two different fields. In addition, the WEATHER TYPES category can contain more than one number per day if several types of weather occurred in the same day. For example, the climatology data from Stapleton International Airport for the month of July 1990 reported July 9 had heavy fog, thunderstorms, hail, and haze. You need to determine the maximum number of weather types that could occur on a single day in order to know how many fields to reserve for this data. For this example, use six fields for weather types. Thus the number of fields in each record is 21 plus 2 for degree days plus 6 for weather types for a total of 29 fields. (You do not need to add a record number to the database, because the day of the month also serves as a record number.)

Types of Data

A database field can contain numeric values or labels but not both. As you look at the data in Figure 5-3, you can see that most of the information is numeric. Some fields, such as maximum temperature, include an asterisk to identify the extreme for the month. You do not need to include the asterisk because it will be easy to select the extremes using the database functions that are discussed later in this chapter.

The precipitation fields can contain T, for trace amounts, but the field typically contains values that enable you to perform numeric operations. Therefore, you need to decide how to handle the trace amounts. For this worksheet, choose –0.01 to represent trace amounts. By using a negative value, you can easily select these fields and ignore them when computing sums and averages. If you used a very small positive value, you would not be able to distinguish it from an actual measurement. The peak gust direction is the only measurement that is a label (E, SE, NW, and so on); all the other fields contain values.

DEC 1990
DENVER, CO
NAT'L WEA SER OFC
10230 SMITH ROAD

ISSN 0198-7690

LOCAL
CLIMATOLOGICAL DATA
Monthly Summary

STAPLETON INTERNATIONAL AP

LATITUDE 39° 45'N   LONGITUDE 104° 52'W   ELEVATION (GROUND) 5282 FEET   TIME ZONE MOUNTAIN   23062

DENVER, CO   DEC 1990

DATE	TEMPERATURE °F					DEGREE DAYS BASE 65°F		WEATHER TYPES 1 FOG 2 HEAVY FOG 3 THUNDERSTORM 4 ICE PELLETS 5 HAIL 6 GLAZE 7 DUSTSTORM 8 SMOKE, HAZE 9 BLOWING SNOW	SNOW ICE PELLETS OR ICE ON GROUND AT 0500 INCHES	PRECIPITATION		AVERAGE STATION PRESSURE IN INCHES ELEV 5332 FEET ABOVE M.S.L	WIND (M.P.H.)								SUNSHINE		SKY COVER (TENTHS)	
	MAXIMUM	MINIMUM	AVERAGE	DEPARTURE FROM NORMAL	AVERAGE DEW POINT	HEATING (SEASON BEGINS WITH JULY)	COOLING (SEASON BEGINS WITH JAN)			WATER EQUIVALENT (INCHES)	SNOW, ICE PELLETS (INCHES)		RESULTANT DIR.	RESULTANT SPEED	AVERAGE SPEED	PEAK GUST SPEED	PEAK GUST DIRECTION	FASTEST 1-MIN SPEED	FASTEST 1-MIN DIRECTION	MINUTES	PERCENT OF TOTAL POSSIBLE	SUNRISE TO SUNSET	MIDNIGHT TO MIDNIGHT
1	2	3	4	5	6	7A	7B	8	9	10	11	12	13	14	15	16	17	18	19	20	21	22	23
01	40	21	31	-4	22	34	0		0	0.00	0.0	24.760	33	3.8	6.3	23	NW	17	30	532	93	1	1
02	37	16	27	-8	8	38	0		0	0.00	0.0	24.630	31	6.8	8.7	48	W	28	29	342	60	6	3
03	41	13	27	-7	4	38	0		0	0.00	0.0	24.890	16	4.9	6.1	17	S	14	19	570	100	0	0
04	66	29	48	14	14	17	0		0	0.00	0.0	24.780	18	6.7	8.8	20	S	14	18	521	91	8	6
05	55	27	41	7	16	24	0		0	0.01	0.1	24.740	10	4.9	7.4	26	N	16	10	227	40	8	8
06	37	19	28	-6	18	37	0		1	0.04	0.7	24.940	17	2.4	4.6	14	N	12	01	547	96	1	4
07	60	23	42	8	17	23	0		0	0.00	0.0	24.830	18	5.5	6.7	18	S	13	18	499	88	4	2
08	64	22	43	9	15	22	0		0	0.00	0.0	24.820	17	2.9	4.4	14	S	9	19	526	93	0	0
09	62	28	45	11	17	20	0		0	0.00	0.0	24.870	18	5.6	6.0	14	S	9	19	531	94	0	0
10	68	35	52	18	17	13	0		0	0.00	0.0	24.730	18	7.1	7.7	17	S	15	20	542	96	0	0
11	68*	37	53*	20	14	12	0		0	0.00	0.0	24.510	26	6.3	10.2	24	W	18	31	544	96	0	1
12	52	32	42	9	24	23	0		0	T	T	24.640	11	7.3	8.6	28	E	15	08	440	78	8	7
13	39	24	32	-1	27	33	0	1 8	0	0.00	0.0	24.610	34	4.3	6.4	15	NW	9	34	361	64	7	8
14	44	21	33	0	9	32	0	2 8	0	T	T	24.440	29	13.6	15.1	51	W	33	29	463	82	4	4
15	50	18	34	1	0	31	0		0	0.00	0.0	24.620	24	0.7	5.0	31	NW	18	27	532	94	1	1
16	50	22	36	3	10	29	0		0	0.00	0.0	24.475	03	1.8	3.3	16	NE	7	04	411	73	6	8
17	37	24	31	-2	21	34	0		0	0.06	0.6	24.450	19	1.2	5.4	17	SE	8	19	460	82	6	5
18	46	25	36	4	12	29	0		0	0.00	0.0	24.260	26	3.4	11.0	40	W	28	28	412	73	6	6
19	28	-3	13	-19	5	52	0	1	0	0.05	1.5	24.360	04	7.6	9.2	22	N	18	01	26	5	10	8
20	-3	-16	-10	-42	-16	75	0	2 8	1	0.06	1.0	24.360	04	8.4	9.8	18	N	15	01	0	0	10	10
21	-7	-21	-14	-46	-22	79	0		2	0.01	0.2	24.540	07	3.9	4.8	13	E	9	05	279	50	4	5
22	-3	-25*	-14*	-46	-20	79	0	1 8	2	0.00	0.0	24.510	36	0.6	3.2	10	SW	7	35	456	81	1	1
23	21	-17	2	-30	-11	63	0	1	2	0.00	0.0	24.720	17	4.0	5.7	18	S	14	19	536	96	0	0
24	37	6	22	-9	1	43	0		2	0.00	0.0	24.640	18	2.2	6.0	20	S	15	18	499	89	7	4
25	18	2	10	-21	4	55	0		2	0.00	0.0	24.680	28	1.3	4.7	14	NW	9	33	455	81	7	5
26	29	1	15	-16	8	50	0		1	0.00	0.0	24.670	14	1.2	3.6	13	E	8	13	452	80	7	5
27	53	4	29	-2	8	36	0		1	0.00	0.0	24.440	20	3.9	5.9	22	S	14	19	542	96	4	3
28	36	2	19	-12	9	46	0	1	1	0.01	0.1	24.310	07	5.5	7.1	23	NE	15	05	442	79	5	6
29	2	-7	-3	-34	-8	68	0	1	1	0.03	0.5	24.530	05	5.1	7.8	17	NE	12	05	123	22	9	10
30	41	-8	17	-13	3	48	0		1	0.00	0.0	24.530	23	2.3	6.4	23	SE	14	33	549	97	2	1
31	57	16	37	7	10	28	0		1	0.00	0.0	24.740	23	3.1	8.0	29	NW	16	31	542	96	2	2

	SUM	SUM				TOTAL	TOTAL			TOTAL	TOTAL			FOR THE MONTH:					TOTAL	%	SUM	SUM
	1225	370				1211	0	NUMBER OF DAYS		0.27	4.7	24.620	17	0.6	6.9	51	W	33	29	13361		134	124
	AVG.	AVG.	AVG.	DEP.	AVG.	DEP.	DEP.	PRECIPITATION		DEP.						DATE:14		DATE: 14		POSSIBLE	MONTH	AVG.	AVG.
	39.5	11.9	25.7	-6.9	7.3	207	0	≥ .01 INCH. 8		-0.28										17505	76	4.3	4.0

NUMBER OF DAYS				SEASON TO DATE		SNOW, ICE PELLETS		GREATEST IN 24 HOURS AND DATES		GREATEST DEPTH ON GROUND OF SNOW, ICE PELLETS OR ICE AND DATE	
				TOTAL	TOTAL	≥ 1.0 INCH 2					
MAXIMUM TEMP.		MINIMUM TEMP.		2301	782	THUNDERSTORMS 0		PRECIPITATION	SNOW, ICE PELLETS		
≥ 90°	≤ 32°	≤ 32°	≤ 0°	DEP.	DEP.	HEAVY FOG 2		0.09 19-20	1.8 19-20	2 24+	
0	8	29	7	-41	102	CLEAR 12 PARTLY CLOUDY 13 CLOUDY 6					

* EXTREME FOR THE MONTH - LAST OCCURRENCE IF MORE THAN ONE.
T TRACE AMOUNT.
+ ALSO ON EARLIER DATE(S).
HEAVY FOG: VISIBILITY 1/4 MILE OR LESS.
BLANK ENTRIES DENOTE MISSING OR UNREPORTED DATA.

DATA IN COLS 6 AND 12-15 ARE BASED ON 21 OR MORE OBSERVATIONS AT HOURLY INTERVALS. RESULTANT WIND IS THE VECTOR SUM OF WIND SPEEDS AND DIRECTIONS DIVIDED BY THE NUMBER OF OBSERVATIONS. COLS 16 & 17: PEAK GUST - HIGHEST INSTANTANEOUS WIND SPEED. ONE OF TWO WIND SPEEDS IS GIVEN UNDER COLS 18 & 19: FASTEST MILE - HIGHEST RECORDED SPEED FOR WHICH A MILE OF WIND PASSES STATION (DIRECTION IN COMPASS POINTS). FASTEST OBSERVED ONE MINUTE WIND - HIGHEST ONE MINUTE SPEED (DIRECTION IN TENS OF DEGREES). ERRORS WILL BE CORRECTED IN SUBSEQUENT PUBLICATIONS.

noaa

NATIONAL OCEANIC AND ATMOSPHERIC ADMINISTRATION

NATIONAL ENVIRONMENTAL SATELLITE, DATA AND INFORMATION SERVICE

NATIONAL CLIMATIC DATA CENTER ASHEVILLE NORTH CAROLINA

Kenneth D Hadler
DIRECTOR
NATIONAL CLIMATIC DATA CENTER

Figure 5-3 **Climatology Data for Stapleton International Airport**

Field Names

It is very important that you use field names that make it easy to identify the fields. For this list, which has been entered in the DEC90 worksheet included in the data files for this module, the following field names were entered across row 10, beginning with cell A10 and ending with cell AC10:

1. DATE	16. WATER
2. MAXIMUM	17. SNOW
3. MINIMUM	18. PRESSURE
4. AVERAGE	19. WINDDIR
5. DEPARTURE	20. WINDSPEED
6. DEWPOINT	21. WINDAVE
7. HEATDAYS	22. PKSPEED
8. COOLDAYS	23. PKDIR
9. TYPEA	24. MINSPEED
10. TYPEB	25. MINDIR
11. TYPEC	26. SUN
12. TYPED	27. SUN%
13. TYPEE	28. DAYCOVER
14. TYPEF	29. NIGHTCOVER
15. COVER	

Also, define the following range names for use with the list:

DATA   A11:AC41

INPUT   A10:AC41

Figure 5-4 shows a window of this list. In the next sections of this chapter, you will use this list, along with the DEFECT list presented in the first section of this chapter, to explore the various operations and computations that can be performed with Excel lists. The data files provided for this module contain 11 additional worksheets with weather information, so a full 12 months of data are available for use with the end-of-chapter exercises.

What If

What values are calculated by the following references? Check your answer using the worksheet.

♦ =MAX(B11:B41)

♦ =MIN(B11:B41)

♦ =SUM(Q11:Q41)

♦ =AVERAGE(Q11:Q41)

♦ =INT(R11)

Give range names to B11:B41, Q11..Q41, and R11. Then modify the previous function references to use the range names.

Figure 5-4
DEC90 Worksheet

	A	B	C	D	E	F	G	H
				TITLE:				
1	TITLE:		Climatology Data for December 1990					
2								
3	DESCRIPTION:		This database contains the climatology data					
4			collected at Stapleton International Airport					
5			for the month of December 1990.					
6								
7	FIELDS:		See the CLIMATE worksheet for a					
8			description of the fields and their codes.					
9								
10	DATE	MAXIMUM	MINIMUM	AVERAGE	DEPARTURE	DEWPOINT	HEATDAYS	COOLD
11	1	40	21	31	-4	22	34	
12	2	37	16	27	-8	8	38	
13	3	41	13	27	-7	4	38	
14	4	66	29	48	14	14	17	
15	5	55	27	41	7	16	24	
16	6	37	19	28	-6	18	37	
17	7	60	23	42	8	17	23	
18	8	64	22	43	9	15	22	
19	9	62	28	45	11	17	20	
20	10	68	35	52	18	17	13	

5-2   SORTING A LIST

Since the information in a list is often used for different reports and analyses, you will want to reorder the records accordingly. While the Data Sort command can be used with any data, it is most commonly used to sort records in lists. The first step is to select a data range that corresponds to the records that you want to sort, which are typically all the records in the list (but not the field names). If you choose the Data Sort command, the Sort dialog box appears as shown in Figure 5-5.

Figure 5-5
Sort Dialog Box

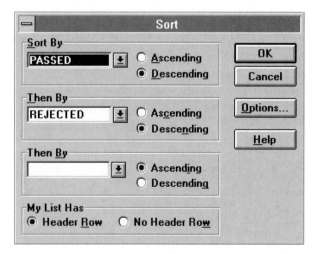

Sort Keys

The sort order is determined by a field you select called the *sort key*. You can use up to three sort keys at a time. (If you need to sort by four or more fields, you must sort the list multiple times.) The first sort key you select is called the *primary sort key*. If the primary sort key is a numeric field, a descending sort reorders the records so that the order is from high to low. An ascending sort reorders the records so that the order is from low to high. If the primary sort key contains labels, an ascending sort reorders the records so that the key column is in alphabetical order; a descending sort is in the reverse order. If the sort key is a mix of numbers and text, you should format all the cells as text; otherwise, Excel sorts the numbers first and then the text separately. To format numbers as text, precede the number with an apostrophe ('). When the key column is all numbers, the numbers should all be in a numeric format.

If necessary, you can select additional sort keys to determine the order of records that have the same values for the previous sort keys. For example, if the primary sort key is a field containing a last name, and the *secondary sort key* is a field containing a first name, then an ascending sort is alphabetical on the last name, and for records with the same last name, the order is alphabetical on the first name.

To select the primary sort key, click the Sort By box in the Sort dialog box, and then type the name of the field that you want to sort. Specify the sort order by clicking the Ascending or Descending button. If you want to select a secondary sort key, select the Then By box and repeat the steps you used when you selected the primary sort key. Be sure to click the appropriate option button for Header Row or No Header Row. To sort the list, click OK. To view the list in the new order, you may want to use the Window Freeze Panes command to freeze the field names at the top of the screen for easier viewing.

Sorting Orders

If the primary sort key field contains labels, the reordering of the records depends on the country format for letters, numbers, and special characters that was selected when Excel was installed. For the United States, Microsoft Excel has the sorting sequences shown in Table 5-2, which contains the ascending (low to high) order. Note that Excel does not distinguish between uppercase and lowercase letters. You can change the sorting sequence using the Tools Options Custom Lists command. Use online Help to find out how to create a custom sorting order.

Table 5-2   Sorting Orders

Type of Data	Sort Order (Low to High)
Numbers	From smallest negative number to largest positive number
Dates and times	From earliest to most recent
Special characters	space,!"#$%&'()*+,-./:;<=>?@[\]^_`{|}~
Text	A to Z, not case sensitive
Logical values	FALSE, TRUE
Error values	All error values evaluate the same (these start with #)
Blanks	Always last, even when sorting from high to low

To demonstrate sorting, you now will examine some sorting examples with the DEFECT list. The data has been entered in chronological order, with the most recent dates last. You probably want to keep this order for the list, so copy the list to a new position in the worksheet; you can experiment with the Data Sort command without worrying about losing your data. Select the list, and use the Edit Copy command to copy the list to the clipboard. Then go to cell A60 and use the Edit Paste command. This puts a copy of the selection with cell A60 as the upper-left corner. Define new ranges named INPUT1 and DATA1 to refer to the copy of the list:

INPUT1   A60:I96

DATA1   A61:I96

Assume that you are interested in the production of boards that pass inspection. You want to sort the list such that the order starts with the month in which the most boards passed inspection and ends with the month in which the fewest boards passed inspection. Therefore, the primary sort key should be the PASSED field (column D), and the sort should be in descending order. For the secondary sort key, select a cell under the RECORD field (column A), and indicate a descending sort. Then, if there is more than one month with the same number of boards passing inspection, they will be listed with the most recent months first. Select the DATA1 range, and then use the following sort settings:

Sort by   PASSED, Descending

Then by   RECORD, Descending

Click the OK button, and your sheet should look like the sorted list shown in Figure 5-6. Note that the record number can be useful as a sort key because you can return the list to the original order using an ascending sort on the record number. You also can use the Edit Undo Sort command to return to the original order before you perform another sort.

Suppose that you are interested in comparing the defect information for January for the three years, then for February for the three years, and so on. To sort the list in this order, you might assume that you want to use the month as the primary sort key (ascending) and the year as the secondary sort key (ascending). However, since the month is stored as a label, the sort on the month would be an alphabetical sort, and the order of the months would change. For example, the first three records would be those for April 89, April 90, and April 91. To solve this problem, you need to store the date information using numbers. If the month is a value from 1 to 12, then using the month field as the primary sort key yields the three records for January as the first three records.

If you sort ranges that contain formulas, the relative cell addresses in a formula are adjusted to reflect the new position of the cell, but absolute addresses are not modified.

There may be applications in which you want to sort only part of the list. For example, using the DEFECT list, you might want to sort and print only records with more than ten defects. The next section discusses the

Figure 5-6
**List Sorted by
Number of Boards
Passing Inspection**

	A16	⬇		RECORD					
	A	B	C	D	E	F	G	H	I
16	RECORD	YEAR	MONTH	PASSED	REJECTED	BOARD	CHIP	PROCESS	SOLDER
17	11	89	Nov	1380	34	20	11	0	3
18	1	89	Jan	1201	34	8	12	4	10
19	10	89	Oct	1180	21	11	6	1	3
20	13	90	Jan	1151	25	6	9	2	8
21	9	89	Sep	1075	18	10	3	4	1
22	21	90	Sep	1071	21	13	3	4	1
23	4	89	Apr	1022	18	9	3	3	3
24	35	91	Nov	1015	14	8	3	0	3
25	27	91	Mar	1010	23	12	6	0	5
26	16	90	Apr	1010	14	8	2	3	1
27	17	90	May	1002	9	3	3	1	2
28	26	91	Feb	989	17	5	5	2	5
29	28	91	Apr	983	19	10	3	3	3
30	19	90	Jul	983	18	7	5	2	4
31	5	89	May	975	10	4	3	1	2
32	14	90	Feb	972	13	3	4	2	4
33	20	90	Aug	956	16	6	9	0	1
34	34	91	Oct	952	18	8	6	1	3
35	3	89	Mar	933	24	13	6	0	5
36	33	91	Sep	918	14	10	1	2	1
37	25	91	Jan	910	18	4	2	4	8
38	18	90	Jun	907	12	8	3	0	1

Data Filter command, which allows you to extract information from a list and store it in another part of the worksheet. You can then sort and print the extracted information.

Try It Copy the DEC90 database to a new worksheet and sort it in the order requested:

◆ By days with the most sunshine first (there are two ways to specify this sort)

◆ By minimum temperature with coldest days first

◆ By maximum temperature with warmest days first, and for days with the same maximum temperature, the one with the least daytime sky cover first

5-3   UPDATING A LIST

Once you have created a list, you will probably need to work with it in several ways. For example, you may need to insert new information (either new records or new fields), update information, or delete records. You may want to select certain records from the list for analysis, for printing, or for creating new lists. If you want to manipulate the data in a list without changing the original list, you can work with a filtered list, which is a copy of selected data from the list. The next section begins the discussion of updating a list with techniques for inserting new fields and new records.

Inserting New Fields and New Records

It is generally best to add, or append, new records at the end of the list and then use the Data Sort command to reorder the records. Append new records to the list by typing them at the bottom.

To insert a new field at the end of each record, enter a field name in the top row in the column immediately to the right of the last field. Then enter the data in the field for each record.

If you insert a record or field within the list (as opposed to the end of the list), you have to be careful about affecting formulas within the list. Also, remember that if you add a field to a record, it should be added to all the records in the list.

You will need to expand the INPUT and DATA ranges to reflect the new range of the list. Do this by selecting the Insert Name Define command, and editing the range in the Refers To text box. Whenever you add records or fields, you may want to sort the list to the order that is most useful, and then you can use the Fill Handle to enter new record numbers that are in ascending order based on the results of the Data Sort command.

Filtering Lists

Excel's Data Filter command allows you to work with a subset of the records based on various criteria for each field. Filtering displays only the rows that meet a set of criteria or contain a certain value. You can use the AutoFilter command to specify simple criteria on a field-by-field basis. This command is a toggle command: The first time you select it, it is turned on, and the next time you select it, it is turned off. To use the AutoFilter command, you first select the list. Then you choose the Data Filter AutoFilter command. Drop-down list arrows appear in each field name cell. Click the arrow to display a drop-down list, and then select values (criteria) from the list to filter. When you choose criteria from these drop-down lists, only the records that match will be displayed; that is, the other records are hidden. To show the entire list again, either toggle the AutoFilter, or choose the Data Filter Show All command.

Figure 5-7
Custom AutoFilter Dialog Box

Drop-down list

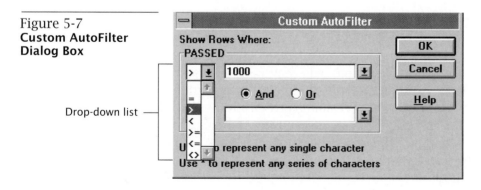

If your criteria for any field is more complex than a simple match, you need to select the Custom option from the drop-down list. The Custom AutoFilter dialog box is shown in Figure 5-7 for the field PASSED. Use the drop-down list to select a logical operator and specify a value.

For additional filter options, you can use an Advanced Filter with a filtering criteria range. A *criteria range* is a range in the same worksheet as an Excel list. A criteria range has fields, field names, and records, and you can determine which records to include. You apply the criteria in the criteria range to your data so that the filtering is done on the original list, or so that the filtered data is copied to another location on the worksheet. If you use another location for the filtered list, you can style and reorganize the data in the new list as you want it to appear without changing the original list.

You begin creating a criteria range in a new part of the worksheet by typing in the field names that you want to use in your filter. These field names must be labels. In the next row, under the field names, type the criteria that you want to use to filter the data. An example of a criteria range is shown in Figure 5-8, cells A54 and A55. Select any cell in the list; Excel chooses the whole list by default. Then choose the Data Filter Advanced Filter command. The Advanced Filter dialog box shown in Figure 5-9 appears. Type the range where the criteria range was created, and edit the list range if necessary. If you would like to copy the results to another part of the worksheet, select the Copy action button. Note that the new location should not overlap or touch the original list. If you do not choose the Copy option, the Advanced Filter will just hide the records that do not match, just as the AutoFilter did. To filter the list, click the OK button.

Figure 5-8
List with Criteria Range

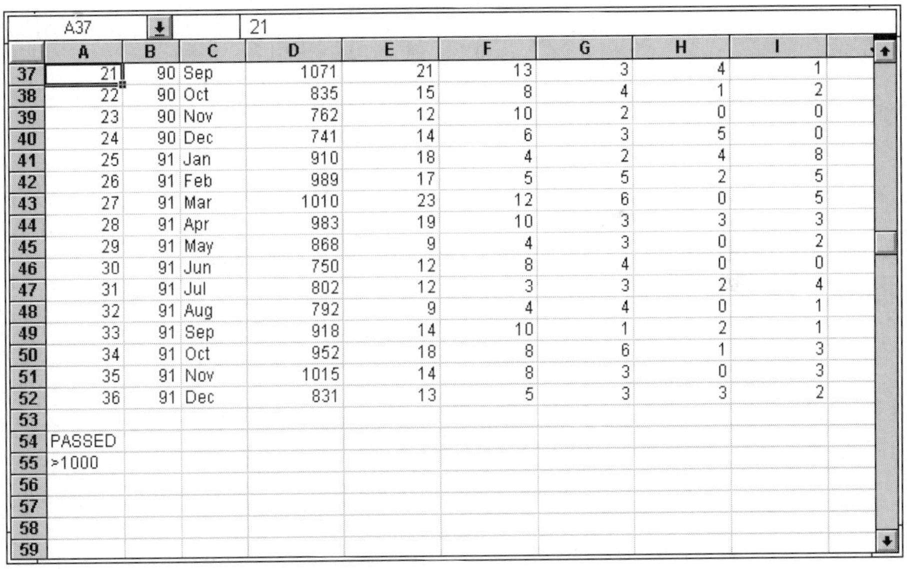

	A	B	C	D	E	F	G	H	I	
	A37	↓		21						
37	21	90	Sep	1071	21	13	3	4	1	
38	22	90	Oct	835	15	8	4	1	2	
39	23	90	Nov	762	12	10	2	0	0	
40	24	90	Dec	741	14	6	3	5	0	
41	25	91	Jan	910	18	4	2	4	8	
42	26	91	Feb	989	17	5	5	2	5	
43	27	91	Mar	1010	23	12	6	0	5	
44	28	91	Apr	983	19	10	3	3	3	
45	29	91	May	868	9	4	3	0	2	
46	30	91	Jun	750	12	8	4	0	0	
47	31	91	Jul	802	12	3	3	2	4	
48	32	91	Aug	792	9	4	4	0	1	
49	33	91	Sep	918	14	10	1	2	1	
50	34	91	Oct	952	18	8	6	1	3	
51	35	91	Nov	1015	14	8	3	0	3	
52	36	91	Dec	831	13	5	3	3	2	
53										
54	PASSED									
55	>1000									
56										
57										
58										
59										

To summarize, when you set criteria for a filter, the records displayed are only the ones that match the criteria. To show the original list after it has been filtered, use the Data Show All command. When you use the AutoFilter command, you choose the criteria by using the drop-down lists. When you use the Custom option, you select the criteria in the Custom AutoFilter dialog box. When you use the Advanced Filter command, you set the criteria based on the values in the criteria range. To create the criteria range, you first enter the field names you are interested in. Then in

Figure 5-9
**Advanced Filter
Dialog Box**

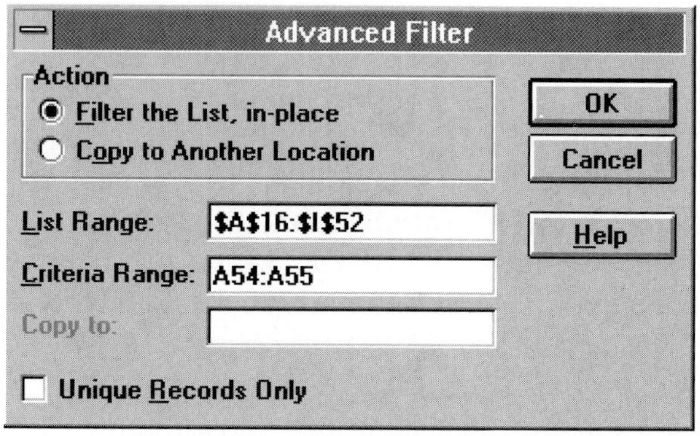

each field you specify the criteria. Criteria can contain numbers, labels, formulas, functions, and logical operators.

The following illustrates the filtering process with a series of examples of the different types of criteria that can be used with the DEFECT and DEC90 lists. These examples begin with criteria to match labels in the list. (Recall that the DEFECT list contains three-character labels in the MONTH field, and that the DEC90 list contains one- or two-character labels in the PKDIR or peak wind direction field.)

Match a specific label: To select all records in the DEFECT list from the month of January, use the AutoFilter command. This is the easiest way to apply this simple filter criteria. Click any cell on the DEFECT list, and select Data Filter AutoFilter. Select the Month drop-down list, and then select the Jan list item. To restore the original list, toggle the AutoFilter command to turn it off.

Match labels with one or more different characters: A question mark (?) can be used in a label to specify a *wildcard* character, which will match any single character. Because June and July are abbreviated to Jun and Jul, records containing the month of June or July can be selected by using the Custom option from the Month drop-down list. In the Custom AutoFilter dialog box, you want to select Show Rows Where: MONTH = Ju?.

Match labels that begin with the same characters but may end differently: Another wildcard character is the asterisk (*). The asterisk is similar to the question mark, but it can represent any number of characters, while a question mark represents a single character. Thus, to match all months beginning with J, you can type either J* or J?? in the Custom AutoFilter dialog box. This criteria selects records from January, June, and July.

Match all labels that meet a specific condition: Chapter 3 discussed logical operators (=, <, <=, >, >=, <>) and their use in comparing numeric values. These same operators can be used with labels to compare an alphabetical order. To select the records that contain the month of September or October, use the drop-down list in the Custom AutoFilter dialog box. Select Show Rows Where: MONTH is >= Oct, as shown in Figure 5-10.

Figure 5-10
Matching Labels with the Custom AutoFilter Dialog Box

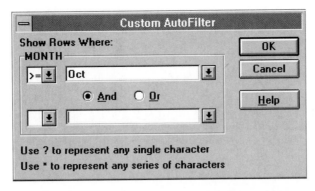

Match all labels except the one specified: You use the not-equal-to operator (<>) to match all records except those matching a specified label. Thus, if you want to select all records from the DEC90 list except those with peak wind gusts from the north, use the Custom AutoFilter dialog box, selecting Show Rows Where: PKDIR <>N.

Match the label that looks like a number: The fields with labels in our two example lists do not contain labels that look like numbers. However, suppose that a list contained phone numbers that were stored as labels in the following form: 793-9321. The following criteria selects all phone numbers beginning with the 793 prefix: Show Rows Where: PHONE=793*.

The following list presents criteria for selecting numbers, as opposed to labels.

Match a number: The criterion for a number is just the number itself. The format does not need to match the list format. Thus, to select months with no soldering errors from the DEFECT list, use the SOLDER AutoFilter drop-down list value of 0.

To select all the months in the DEFECT list with more than ten board defects, use the Custom AutoFilter dialog box for the field BOARD. Select Show Rows Where: BOARD >10.

Match a number using a formula: This complex criteria requires the use of the Advanced Filter command. Be sure to toggle the AutoFilter off. Now create a criteria range outside of your list. Thus, if you want to select the days from the DEC90 list in which the maximum and minimum temperatures differ by more than 30 degrees, type the following in cells A44 and A45 (which do not overlap the list):

TEMPDIFF

=MAXIMUM>(MINIMUM+30)

Fill in the range for the criteria as shown in Figure 5-11.

Notice that you had to give the criteria range a new field name (TEMPDIFF), and that the value becomes #NAME? when it is displayed. It is okay to ignore this value; the criteria is displayed on the edit line whenever you select this cell.

Figure 5-11
**Advanced Filter
Dialog Box for
Matching a Number
Using a Formula**

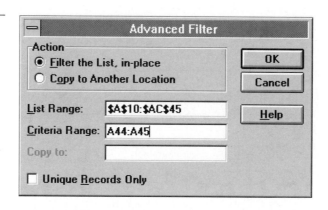

Select any cell on the database list, and then select the Data Filter Advanced Filter command. Excel fills in the list range.

There are 14 days that meet these criteria. To show the original list again, use the Data Filter Show All command.

Match a value using a function: You also can enter a function as a value in a criteria range. Assume you want to locate the days from the DEC90 list with the highest maximum temperature. First use Insert Name Define to name the range containing the maximum temperatures MAXRANGE, and then at C44 and C45 type the following criteria range:

MAXTEMP

=MAXIMUM=MAX(MAXRANGE)

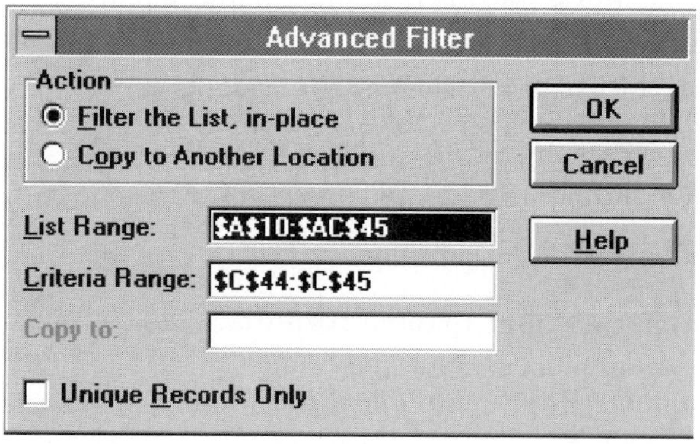

Figure 5-12
**Advanced Filter
Dialog Box for
Matching a Value
Using a Function**

Select a cell in the list, select the Data Filter Advanced Filter, type in the criteria range as shown in Figure 5-12, and then click the OK button.

The maximum temperatures occurred on days 10 and 11.

The previous examples were of criteria to match labels and numbers. Now we examine how to combine these criteria so that you can select records that meet combinations of criteria.

Matching all criteria: With the AutoFilter you can select all the records that meet all of the field criteria. To select fields with up to two criteria, you need to use the Custom option for that field. To select records that meet all the criteria at once in an Advanced Filter, you can put all the criteria on the same row, or you can create a function using the logical function =AND. Either way, by combining the criteria with AND, Excel selects only records that satisfy both criteria. For example, suppose you want records from 1991 with more than 15 rejected printed circuit boards. First, use the AutoFilter command to select 91 from YEAR and >15 from REJECTED. Another way to solve this problem is to use an Advanced Filter by first specifying the following criteria range in cells A55:A56:

MANY_REJECTS

=AND(YEAR=91,REJECTED>15)

Then select any cell in the list, and select the Advanced Filter command. Fill in the dialog box as shown in Figure 5-13:

Figure 5-13
**Advanced Filter
Dialog Box for
Matching All Criteria**

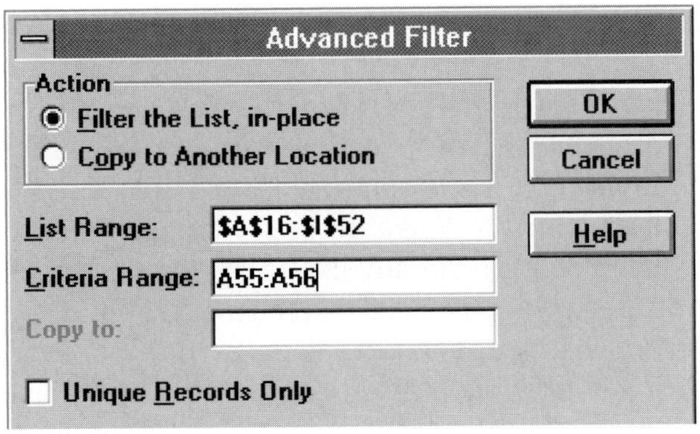

A third solution uses this criteria range in cells E55:E56:

YEAR	REJECTED
91	>15

The same five records are selected by all three filters.

Matching any of the criteria: To select records that meet some criteria but not necessarily all of the criteria, you use the OR logical operator. The AutoFilter only applies OR to a custom criteria on a field name; it will not OR the results of different field criteria. For example, assume you want to select records with more than ten chip defects and either no processing defects or more than eight solder defects. You must use an Advanced Filter. First specify the criteria range in cells A58:A59:

SHOW_DEFECTS

=AND(CHIP>10,OR(PROCESS=0,SOLDER>8))

Another way to do this is to put the items on separate rows of the criteria range. Type the following in E58:G60:

CHIP	PROCESS	SOLDER
>10	=0	
>10		>8

Both of these criteria represent the same values. Click a cell in the list, and select the Advanced Filter command. When the dialog box appears, type either criteria range. The selected records correspond to January 1989 and November 1989.

To change or add new criteria for criteria range, you select the cells to edit or type in additional ones. When you filter a list, be sure to select the correct criteria range, since Excel will remember and use the criteria range from a previous filter as a default.

Working with Filtered Lists   You can perform several operations on a filtered list. If you used the Copy option of the Advanced Filter command, you can name the new filtered list as you would any other range. The values in this range do not change when the original list is updated. You can also perform any list operation on this new filtered list. However, if you used an Advanced Filter without the Copy option, any successive filtering will be applied to the entire original list, not just the filtered list.

If a worksheet contains several new filtered lists, you can name each filtered list using the Insert Name Define command. You can then use the Edit Go To command to locate a specific filtered list. You can also name your criteria ranges. If the original list changes after you create a filtered list, you can refresh the filtered list to reflect the changes by applying the criteria range to the updated original list, specifying the copy option, and giving the range name of the filtered list. Be sure to leave enough space between lists and criteria ranges so that they do not overlap.

Modifying List Records

You can either modify the list records directly or use the Data Form command to find list records that match criteria so you can update them. To use the Data Form command, select any cell in the list. When you choose the Data Form command, the Data Form dialog box shown in Figure 5-14 appears. To set the criteria, click the Criteria button, and type the criteria in the appropriate edit boxes. When you have set the criteria, you can move through the records that match all the criteria using Find Next and Find Previous buttons. To change a field in the selected records, type the new values in the edit boxes and then press (ENTER). Note that computed fields cannot be edited.

Figure 5-14
Data Form Dialog Box

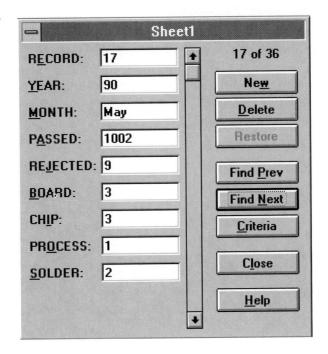

Try It Copy the DEFECT list to an unused area of the worksheet. Create named ranges for the input area and data area. Then perform the following operations:

♦ Update the number of board defects in Nov 1989 to 15. Also, adjust the number of boards rejected accordingly.

♦ Create a filtered list that includes all the database records with more than 15 boards rejected. Print a list of these records.

♦ Create a filtered list that does not include records that have zero process defects or zero solder defects.

♦ Create a filtered list that includes all records with more process defects than solder defects.

5-4 PERFORMING STATISTICAL COMPUTATIONS

This section presents some special features of Excel for performing statistical computations with a list. First, list management functions are presented. These functions are similar to the statistical functions presented in Chapter 3, but they have been modified so that you can use them with records selected from the list using a criteria. Then this section presents the Histogram command, which computes a frequency distribution using information from the list.

Database Functions

Functions that are used with a list are modified versions of a number of the statistical functions discussed in Chapter 3. Each database function has a name that begins with the letter *D*, to signify the database version, and has three arguments: database, field, and criteria. The first argument is the range of the list, the second argument specifies the column of the list to be used, and the third argument specifies the criteria to use. The advantage of these database functions is that you can specify a set of criteria in a criteria range. Database function criteria ranges work like the criteria ranges for an Advanced Filter. The database function applies only to the records that match the criteria. For example, you can count only records with a total number of defects over 15, or you can compute the average number of boards rejected during the first six months of 1990.

A database function has the following arguments:

- The database argument is the range name or range address that contains the list or filtered list.
- The field argument is the field name enclosed in quotation marks.
- The criteria is the range name or address of the criteria range.

Here is a list of the database functions along with a description of the values they compute:

Function	Description
=DCOUNT(*database, "field", criteria*)	Counts the number of values that meet the specified criteria.
=DCOUNTA(*database, "field", criteria*)	Counts the number of non-blank cells that meet the criteria for the given field.
=DMAX(*database, "field", criteria*)	Finds the largest of the set of values in the indicated field that meet the specified criteria.
=DMIN(*database, "field", criteria*)	Finds the smallest of the set of values in the indicated field that meet the specified criteria.
=DAVERAGE(*database, "field", criteria*)	Computes the average of the set of values in the indicated field that meet the specified criteria.
=DSUM(*database, "field", criteria*)	Computes the sum of the set of values in the indicated field that meet the specified criteria.

=DPRODUCT(*database*,*"field"*, *criteria*)	Computes the product of the set of values in the indicated field that meet the specified criteria.
=DGET(*database*,*"field"*, *criteria*)	Extracts a single value from the set of values in the indicated field that meet the specified criteria. If more than one record meets the criteria the error value #NUM! is returned.
=DSTDEV(*database*,*"field"*, *criteria*) or =DSTDEVP(*database*,*"field"*, *criteria*)	Computes the standard deviation of the set of values that meet the specified criteria. (DSTDEV – sample standard deviation; DSTDEVP – population standard deviation)
=DVAR(*database*,*"field"*, *criteria*) or =DVARP(*database*,*"field"*, *criteria*)	Computes the variance of a set of values that meet the specified criteria. (DVAR – sample variance; DVARP – population variance)

The following examples use the DEFECT list. Recall that the list name is INPUT. Set up the criteria range shown in Figure 5-15.

Figure 5-15
Criteria Ranges for the DEFECT Database Examples

	D55	↧		=CHIP>PROCESS					
	A	B	C	D	E	F	G	H	I
41	25	91	Jan	910	18	4	2	4	
42	26	91	Feb	989	17	5	5	2	
43	27	91	Mar	1010	23	12	6	0	
44	28	91	Apr	983	19	10	3	3	
45	29	91	May	868	9	4	3	0	
46	30	91	Jun	750	12	8	4	0	
47	31	91	Jul	802	12	3	3	2	
48	32	91	Aug	792	9	4	4	0	
49	33	91	Sep	918	14	10	1	2	
50	34	91	Oct	952	18	8	6	1	
51	35	91	Nov	1015	14	8	3	0	
52	36	91	Dec	831	13	5	3	3	
53									
54	REJECTED	MONTH	YEAR	C>P					
55	>25	Jan	89	#NAME?					
56			90						
57									
58									

To count the number of months in the list that rejected over 25 boards, use the following function reference:

=DCOUNT(INPUT,"REJECTED",A54:A55)

This function reference computes the value 2. Note that the field name is REJECTED, just as it is in the worksheet.

To compute the average (rounded to the nearest integer) of the number of boards passed during the first month of each year, use the following criteria and function reference:

$$=ROUND(\ AVERAGE(INPUT,``PASSED", \ B54:B55),0)$$

This function reference computes the value 1087.

To compute the sum of all chip errors for 1989 and 1990, use the following criteria:

$$=DSUM(INPUT,``CHIP",C54:C56)$$

This function reference computes the value 119.

The following criteria counts all months in which the number of chip defects exceeded the number of processing defects:

$$=DCOUNT(INPUT,``CHIP",D54:D55)$$

This function reference computes the value 26. Note that the field name "C>P" in the criteria range is not one of the original field names. This is because we are comparing the values of two different fields. Also, the value #NAME? is shown in cell D55 because this formula can only be evaluated as part of a criteria.

The Histogram Tool

The Data Analysis Tools command contains a histogram among its many features. If your Tools menu does not have a Data Analysis option, use the Tools Add-ins command to set up the Analysis ToolPak. Choose the Analysis ToolPak from the list of options, then click the OK button. An hourglass icon appears on the screen while the ToolPak is being loaded. The Data Analysis Histogram tool creates a frequency distribution using the values from a specified range, where a frequency distribution is a count of how many values fall within specified intervals. For example, suppose you want to determine a distribution showing how many days in the DEC90 list had maximum temperatures in these ranges: 0 and below, 1–10, 11–20, 21–30, 31–40, 41–50, 51–60, 61–70, 71–80, 81–90, 91–100, and over 100. This distribution can be computed easily using the Histogram command, which uses a values range and bin values as shown in the Histogram Dialog Box in Figure 5-16. The *input range* is the data that is to be counted for the frequency distribution; in the example, the values range is the field containing the daily temperature maximum values, or B11:B41. The *bin values* are the upper limits of the intervals; in the example, the bin values are 0, 10, 20, 30, 40, 50, 60, 70, 80, 90, and 100, which we assume have been stored in A51:A61. The upper-left corner of the frequency distribution table is the *output range*.

After you specify the range containing the values to be counted, the bin values, and the output range, the Histogram command fills the specified output column with the upper limits of the bins, and the next column to the right with the corresponding counts. The command uses one row past the last bin value to specify the count for values greater than the last

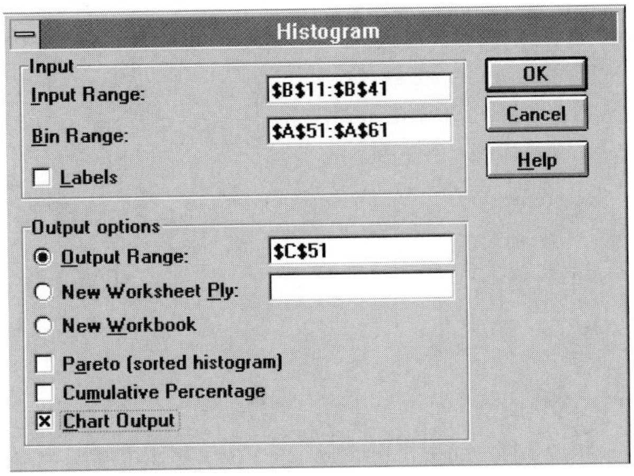

Figure 5-16
Histogram Dialog Box

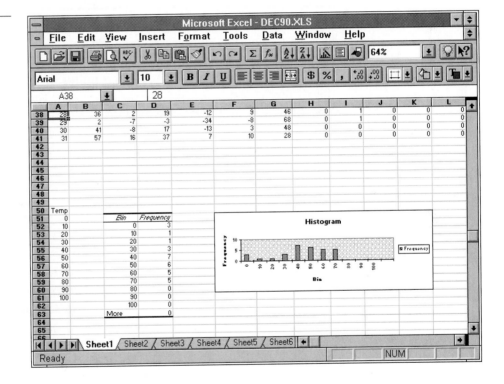

interval. You can even select the chart option that draws the histogram as a chart, as shown in Figure 5-17.

Figure 5-17
Output of Histogram Tool

You could easily change the headings and print this frequency distribution in a report, or edit the chart in any of the ways shown in Chapter 4.

Try It

1. Give the criteria and the database function reference to compute the following using the DEC90 list:

   ◆ Average maximum temperature

   ◆ Average minimum temperature

   ◆ Average average temperature

- ◆ Number of days with heavy fog
- ◆ Percentage of days with snow
- ◆ Peak wind gust

2. Generate a frequency distribution that counts the peak wind gusts in the following intervals, using the DEC90 list:

0–10 mph

11–20 mph

21–30 mph

31–40 mph

41–50 mph

Over 50 mph

3. Add a histogram to the frequency distribution generated in the previous problem.

SUMMARY

This chapter demonstrated how to use the Excel lists as databases. You learned to use this feature to analyze data, using climatology data and circuit board defect data. When defining a list, it is very important to include documenting information at the beginning of the worksheet to describe the different fields within the record. Once the list is created, you can sort, update, and display the data using the data operations. For example, you can sort the list in ascending or descending order based on one or more sort keys. You also can append records to the list or find records you want to modify or delete. If you want to manipulate a subset of the list without changing the original table, you can create a new list by selecting the Copy option when you filter the list. Finally, the chapter showed you how to perform statistical computations on the list.

Key Words

criteria	primary sort key
database	record
field	secondary sort key
field name	sort key
list	wildcard

Exercises

The first set of exercises uses the lists generated in this chapter. Start each exercise with the original list unless otherwise directed. The last set of exercises use the commands presented in this chapter with worksheets developed in earlier chapters.

DEFECT Worksheet

1. Prepare a report that contains the date along with the number of boards passing inspection and the number of boards rejected for each month in which the number of boards passing inspection was greater than 1000.

2. Prepare the report from exercise 1, but order the information so it is in ascending order of boards passing inspection. For months with the same number of boards passing inspection, use a descending order of boards rejected.

3. Add summary information to the report in exercise 1 that contains the average number of boards passing inspection and the average number of boards rejected for the months used in the report.

4. Prepare a frequency distribution and a histogram that counts the number of months from the list with the number of boards passing inspection in the following intervals:

   Under 750

   750-850

   851-950

   951-1050

   1051-1150

   Over 1150

5. Prepare a report based on information from months with over 20 boards rejected. The information should contain the boards rejected in the month and the number of defects in each of the four types. Include summary information that computes the percentage of each type of defect for the group in the report.

DEC90 Worksheet

6. Generate a histogram that compares the number of days with the following maximum temperature intervals:

   Below 0

   0-32

   33-50

   51-60

   61-70

   Over 70

7. Generate a pie chart that gives the percentages that the minimum temperatures fell within the intervals given in exercise 6.

8. The summary at the bottom of the monthly report generated by the National Climatic Data Center includes the following information:

   - Sum of the maximum temperatures
   - Average maximum temperature
   - Sum of the minimum temperatures
   - Average minimum temperature
   - Average temperature
   - Average dew point

Use database functions to compute these values and print them in a short report.

9. Use database functions to compute the following values for days with fog:

   - Average temperature
   - Maximum temperature
   - Minimum temperature

   Print this information in a short report.

10. Generate an *XY* graph containing the maximum and minimum temperatures by day for the month of December 1990.

11. Generate a new worksheet containing the data from January through April 1991. (Use the monthly lists from the data files provided for this module.) Generate a summary report for this period of time that contains the following:

    - Maximum temperature and the date (or dates) it occurred
    - Minimum temperature and the date (or dates) it occurred

UNIFORM Worksheet

12. Sort the uniform random numbers in an ascending order and generate a line plot of the values.

13. Use the Histogram Tool to generate a frequency distribution of the 100 uniform random numbers using the following intervals (*m* represents the mean and *s* represents the standard deviation):

    $< m - 3s$

    $m - 3s$ to $m - 2s$

    $m - 2s$ to $m - s$

    $m - s$ to m

    m to $m + s$

    $m + s$ to $m + 2s$

    $m + 2s$ to $m + 3s$

    $> m + 3s$

    The values generated in each interval should be close in value to each other (except for the first and last values).

Index

'.   *See* Apostrophe
*.   *See* Asterisk
^.   *See* Caret
,.   *See* Comma
/.   *See* Forward slash
-.   *See* Minus sign
( ). *See* Parentheses
π.   *See* Pi
+.   *See* Plus sign
#.   *See* Pound sign
?.   *See* Question mark

absolute reference, 30
absolute value function, 35
active cell, 13
addition, 32
Advanced Filter command, 99
Advanced Filter dialog boxes, 97-
    98, 100-101
aerospace engineering, 25-26
algorithm, 4-5
alignment, 15
alignment tab, 20-21
Analysis ToolPak, 106
AND expression, 44
apostrophe ('), 93
application programs, 2
arguments, 35, 104
arrow keys, 13
ascending order, 93
ASCII files, 78
asterisk (*), 32, 98
AutoFilter command, 96
automatic recalculation, 33
average, 39, 104, 106

bin values, 106-108
bounds, 41

cable cars, 29, 47-49
CABLE worksheet, 48-49
Calculation dialog box, 34

Cancel box, 13, 31
caret (^), 32
cascade menus, 16-17
case sensitivity, 93
cell addresses, 11, 13, 31
    formulas and, 30
    sorting and, 94
    value displayed in, 14
cell pointer, 13
cell references, 30-31
cells
    active, 13
    defined, 11
    selecting range of, 19
charts
    enhancing, 64-66, 72-73, 76
    format selection, 62-63
    naming, 66
    preliminary, 62-63
    printing, 66
    saving, 66
    type selection, 62
    types of, 59
ChartWizard button, 61
ChartWizard dialog boxes, 62-64
circuit boards, 66-71
circular references, 34-35
climatology data, 87-91
clipboard, 22
col_index_num parameter, 44
colors, 72
column charts, 60, 73-75
columns
    in databases, 84
    defined, 11
    manipulating, 21-22
    selecting, 19
    width, 21
comma (,), 35
commands, 16-18
    Advanced Filter command, 99
    AutoFilter command, 96
    categories of, 12
    Copy command, 22, 33

Cut command, 22, 63
data, 84
Data Analysis Tools command,
    106
Data Filter Advanced Filter com-
    mand, 97
Data Filter AutoFilter command,
    96
Data Filter Show All command,
    96
Data Form command, 102-103
Data Sort command, 92
Data Text to columns command,
    79-80
Delete command, 21
File New command, 23
File Open command, 23, 78
File Print command, 22-23
File Save As command, 23
File Save command, 23
Format AutoFormat command,
    74-75
Format cells command, 20
Format column Width com-
    mand, 21
Format Sheet Rename com-
    mand, 66
Format Style command, 20
Help command, 24
Insert Gridlines command, 65
Insert Name command, 40
Insert Name Define command,
    96, 102
Insert New Data command, 72
Insert Title command, 72
macro, 54-55
Paste command, 22, 33, 63
Print Preview command, 23
Show Topics command, 24
Titles command, 64
Tools Add-ins command, 106
Tools Options command, 33
Window Freeze Panes com-
    mand, 86

Window Unfreeze Panes command, 86
composition, 37
computer-aided engineering (CAE), 1
computer, defined, 2
computer engineering, 41-43
contents box. *See* cell addresses
control panel, 12
Copy command, 22, 33
copying
    filtered lists, 97
    formats, 22
    formulas, 33
    information, 22
criteria, 84
    types of, 98-102
criteria argument, 104
criteria range, 97
currency format, 20
Custom AutoFilter dialog box, 96, 99
Cut command, 22, 63

data
    aligning, 20-21
    editing, 15-16
    importing, 77-80
    types of, 89
Data Analysis Tools command, 106
database argument, 104
database functions, 104-106
databases. *See also* lists
    defined, 84
data commands, 84
data files, 77-80
Data Filter Advanced Filter command, 97
Data Filter AutoFilter command, 96
Data Filter Show All command, 96
Data Form command, 102-103
Data Form dialog box, 103
DATA range, 86, 96
data ranges, 62
Data Sort command, 92
Data Text to Columns command, 79-80
date format, 20
DATE function, 46
DEC90 worksheet, 92
decimals, 20
decomposition, 4
defaults
    chart format, 62
    column width, 21
    directory, 23
    file names, 23
DEFECT worksheet, 85-86
Delete command, 21
delimited file, 78
descending order, 93
destination range, 22
dialog boxes, 16-18

how to use, 24
digital filters, 59
directories, 23
division, 32

editing, 15-16
edit line, 12-13
electrical engineering, 41-43
engineering
    aerospace, 25-26
    electrical/computer, 41-43
    environmental, 87-88
    manufacturing, 66-71
    mechanical, 47-54
enhancements
    for charts, 64-66
    for line charts, 76
    for pie charts, 72-73
Enter box, 13, 31
environmental engineering, 87-88
equilibrium, 50-54
errors
    #N/A!, 44-45
    #NUM!, 105
    #REF!, 45
    in typing, 15
    #VALUE!, 45
Excel window, 11
exploding slice, 73
exponential functions, 37
exponentiation, 32

field argument, 104
fields, 84
    inserting, 96
    names for, 84, 91
    number of, 89
file names, 23
File New command, 23
File Open command, 23, 78
File Print command, 22-23
File Save As command, 23
File Save command, 23
Fill Handle, 86-87
FILTER1 worksheet, 61
flowcharts, 4
Format AutoFormat command, 74-75
Format Cells command, 20
Format Cells dialog box, 20-21
Format Column Width command, 21
Format Legend dialog box, 75
Format Painter button, 22
formats, copying, 22
Format Sheet Rename command, 66
Format Style command, 20
formatting
    numbers, 20
    worksheets, 20-21
formula bar, 13, 14, 31
formulas, 30-35

forward slash (/), 32
fraction format, 20
Frankston, Bob, 2
frequency distribution, 106
Fricklin, Dan, 2
functions, 2-3
    composition of, 37
    database, 104-106
    exponential, 37
    logarithmic, 37
    mathematical, 35-38
    nesting of, 37
    rounding, 38
    special, 43-46
    statistical, 39-40
    trigonometric, 36-37
    truncating, 38
    types of, 30
Function Wizard, 31, 35
Function Wizard button, 13
Function Wizard dialog box, 36

GOTO function, 14
graphics software, 1
graphs. *See* charts
gridlines, 65

hand example, 4
hardware, 2
Help command, 24
Histogram dialog box, 107
Histogram tool, 106-108
HLOOKUP function, 44-45

IF function, 43-44
importing data files, 77-80
information gathering, 3
input/output (I/O), 3
INPUT range, 86, 96
input ranges, 106-108
Insert Gridlines command, 65
inserting
    columns, 21
    fields, 96
    records, 96
Insert Name command, 40
Insert Name Define command, 96, 102
Insert New Data command, 72
Insert Title command, 72
integrated package, 2
iteration, 33-34, 50-54

Kapor, Mitch, 2

labels, 72
line charts, 60, 75-77
list management operations, 84
lists, 84
    creating, 84-92

filtering, 96-102
naming filtered, 102
sorting, 92-95
updating, 95-103
location. *See* cells
logarithmic functions, 37
logical operators, 43-44, 98-99
LOOKUP functions, 44-45
lookup_value parameter, 44
Lotus 1-2-3 spreadsheet program, 2

macro commands, 54
manual recalculation, 34
manufacturing engineering, 66-71
mathematical formulas, 30-35
mathematical functions, 30
mean. *See* average
mechanical engineering, 47-54
menu bar, 12
menu system, 16-18
meteorology, 83
microprocessor, 59
Microsoft Excel integrated software, 2
minus sign (–), 32
mixed reference, 31
mode indicator, 16
mouse, 13
moving information, 22
multiplication, 32

#N/A! error, 44-45
name box, 13, 31
naming
charts, 66
fields, 84, 91
files, 23
filtered lists, 102
ranges, 40
worksheets, 66
National Climatic Data Center, 87-88
National Weather Service, 87-88
nesting, 37
nondelimited file, 78
NOT expression, 44
number format, 20
numbers, random, 39
numerical relationships, 75-76
numeric values range, 15
#NUM! error, 105

Online Help, 24
Open File dialog box, 78
opening worksheets, 23
operations, 31-32
OR expression, 44
output range, 106-108

parameters, 35

parentheses (), 32
parsing, 79-80
Paste command, 22, 33, 63
percentage format, 20
pi, („), 35
pie charts, 60, 71-73
PLATE worksheet, 51-54, 56
plotting rules, 71
plus sign (+), 32
pound sign (#), 20
primary sort key, 93
Print dialog box, 22-23
printing
charts, 66
worksheets, 22-23
Print Preview command, 23
problem-solving process, 3-5
problem statement, 3
program, defined, 2
programming languages, 2, 4
pull-down menus, 16-17

quality control, 66-71
QUALITY worksheet, 67-71
question mark (?), 98

RAND function, 41-43
random numbers, 39
generation, 41-43
range address, 19
range names, 13, 40
ranges
changing for charts, 64
defined, 19
destination, 22
selecting multiple, 19
recalculation, 33-34
Record New Macro dialog box, 54
records, 84
displaying numbers, 86-87
inserting, 96
#REF! error, 45
relative reference, 30-31
rocket stages, 25-26
rounding functions, 38
rows
in databases, 84
defined, 11
manipulating, 21-22
selecting, 19

Save As dialog box, 23
saving
charts, 66
worksheets, 23
scatter charts. *See* XY (scatter) charts
scientific format, 20
Search dialog box, 24
secondary sort key, 93
selecting ranges, 19
sheet tabs, 11

shortcut menus, 17
Show Topics command, 24
software, 2
solution
generation and evaluation of, 4
refining and implementing of, 4-5
testing of, 5
solution template, 4-5
Sort dialog box, 92
sorting sequence, 93
sort keys, 92-93
sounding rockets, 9, 25-26
spaces, 35
special functions, 30, 43-46
spreadsheets. *See* worksheets
square root function, 35
stacked column charts, 60, 74-75
standard deviation, 39-40, 105
Stapleton International Airport, 88-91
statistical functions, 30, 38-40
lists and, 103-108
status bar, 13
stepwise refinement, 4
subtractions, 32
summation, 36, 104, 106

table_array parameter, 44
telemetry systems, 9
temperature distribution, 50-54
text, 15-16
text files, 78
Text Import Wizard, 78-79
3-D column chart, 74
time format, 20
TIME function, 46
title bar, 12
Titles command, 64
Titles dialog box, 65
titles, viewing, 86
toolbar, 12
toolbar buttons, 17-18
formatting with, 22
Tools Add-ins command, 106
Tools Options command, 33
top-down design, 4
trajectory, 9
TRAJECTORY worksheet, 25-26
truncating functions, 38
.TXT extension, 78

UNIFORM worksheet, 41-43

#VALUE!, 45
values, 14-15
raising, 32
variance, 39-40, 105
velocity, 29, 47-49
viewing titles, 86
VisiCalc spreadsheet program, 2
VLOOKUP function, 44-45

weather information, 83
wildcard characters, 98
Window Freeze Panes command, 86
windows, 11-13
Window Unfreeze Panes command, 86
workbooks, 11
worksheets, 4
   blank, 10
   CABLE, 48-49
   DEC90, 92
   defined, 2
   entering text into, 15-16
   entering values into, 14-15
   FILTER1, 61
   formatting, 20-21
   moving within, 13-14
   opening, 23
   operations, 19-23
   PLATE, 51-54
   printing, 22-23
   QUALITY, 67-71
   saving, 23
   TRAJECTORY, 25-26
   UNIFORM, 41-43
   using, 10-16
   window for, 13

.XLS extension, 23
XY (scatter) charts, 60-66

NOTES

NOTES

NOTES

NOTES

NOTES

NOTES